A Colour Handbook

Landscape Gardening

A Colour Handbook
Landscape Gardening

Alka Singh
Associate Professor
Department of Floriculture & Landscape Architecture
ACHF, Navsari Agricultural University
Navsari, Gujarat

B K Dhaduk
Retd. Professor and Head
Department of Floriculture & Landscape Architecture
Retd. Dean, ASPEE Institute of Agri Business Management
ACHF, Navsari Agricultural University
Navsari, Gujarat

NEW INDIA PUBLISHING AGENCY
New Delhi – 110 034

NEW INDIA PUBLISHING AGENCY
101, Vikas Surya Plaza, CU Block, LSC Market
Pitam Pura, New Delhi 110 034, India
Phone: + 91 (11)27 34 17 17 Fax: + 91(11) 27 34 16 16
Email: info@nipabooks.com
Web: www.nipabooks.com

Feedback at feedbacks@nipabooks.com

ISBN: 978-93-83305-88-9

Composed and Designed by NIPA

Preface

The villages are converting into small towns and small towns into metro cities creating concrete jungle with the shrinking of natural vegetation and the developing industrial epoch. Rapid urbanization has been accompanied by expansion of highly vulnerable communities living in informal settlements far away from nature, creating high environmental pressures with poor air quality in the outdoors as well as indoors. A high proportion of the population and economic activities are at risk from climate change, generating a high amount of global greenhouse gas emissions adding to the rise in global warming. Deaths due to air pollution have increased fourfold across the globe over the past decade as per the assessment by the World Health Organization. The accelerated pace of life has made man more mechanical that has further induced everyday stress and health problems.

Landscape Gardening is the ray of a green hope in this fast mechanical era and the *modus operandi* for restoring balance with ecology. It bestows vitality, quality and aliveness to this expeditious and hasty life. A beautifully designed landscape and well planned avenue plantation not only improves aesthetic view but also curbs pollution, provides pure air, modulates temperature, sunlight and wind velocity, besides offering natural habitat for birds and connects human life with nature. A sight of blooming flowers captivates human heart with its enthralling beauty and further stimulates creativity and improves work performance. In the scorching summers, a sight of shady tree and a glimpse of tree blossoms changes the whole ambience.

Sun-Hot and Red
Stepped on Earth with grace
Oozing with love and delight
Smiling in blossoms of Gulmohar!

Interaction with gardens and natural spaces offers a variety of mental, physical and social beneficial influence for humans, ranging from stress reduction, quicker healing, besides mitigating environmental problems. Flowers convey our inner feelings sympathy, romance or celebration (pride and joy) or contrition in the most beautiful and delicate manner. Flowers are also a good natural source for stress busting. Studies throughout the world have proven the power of flowers and green spaces to improve human health. Recent studies in the Netherlands and Japan disclosed that people with easy access to green space were blessed with better health and lower mortality rate. Thus, the great architect Corbusier has rightly quoted: "Unless the conditions of nature are re-established in man's life, he cannot be healthy in body and spirit."

A Glimpse of tree laden with blossoms fills up the heart with immense joy and over flowing love as depicted through this poem on Amaltas (*Cassia fistula*)

Amaltas Blossoms –Captivating Treat

Lamps of glowing joy with the sprinkle of freshness
Shines and shimmers in hanging blossoms of *Amaltas*
Reflects the light of wisdom
With fragrance of throbbing love
Diffuses the fragrance of joy and juvenility
With the loving strokes of tenderness and softness
Captivates the eyes and holds the heart
With the touch of endearing and radiating grace
Sprawling the true feel of exquisite beauty
With the charisma of blissful life!

It is the application of the art and technique of Landscape Gardening that can spell magic to our surroundings enhancing aesthetic beauty besides, improving and conserving environment. It is an art of alluring the glory of nature at your door step. It is the technique to beautify our surroundings, our homes, office buildings and in the open space with plants and gardens features to give the soothing effect on the body, mind and soul. It is an art and science which deals with beautiful elements of the nature, i.e., plants, water and landform with the implementation of components of garden like lawn, shrubberies, hedges, edges, carpet bedding, flower beds, arches, pergolas etc. on a landscape in an aesthetic and rational manner. This book deals with the basic aspects of landscape gardening including garden types and styles, principles, elements and garden components in detail and is well supported with illustrations and photographs for better clarity and easy understanding. It has been compiled especially for the students and it would also be useful to the landscape professionals, teachers as well as garden and nature lovers.

Contents

Preface v

1. Landscape Designing and Gardening: Significance and Scope 1
Definition, Significance and Importance 1
Gardening and Landscaping 2
Importance and Benefits 2
Human Psychology and Health 3
Society and Culture 4
Gardens and Children 5
Other Benefits 5
Professional Scope 6

2. Basic Garden Styles and Conventional Gardens 9
Formal Garden Style 9
Egyptian Gardens 10
Persian Gardens 10
Mughal Gardens 12
Greek/Hellenistic Gardens 13
Roman Gardens 14
The Italian Renaissance Gardens 15
The French Renaissance Gardens 16
Informal Garden Style 17
Chinese Gardens 17
Japanese Gardens 18
Intermediate Garden Style 22
English Gardens 23
Wild Garden Style 24

3. Garden Types and Designs: Contemporary View 27
The Cottage Garden 27
Family Garden 28
Secluded Garden 29
Terrace Garden 31
Paved Garden 32
Roof Garden 32
Rock Garden 35
Moon Garden 37
Bog Garden 38
Sacred Gardens 38
Home Garden 40
Kitchen Garden 41
Herbal Garden 44

4. Garden Designing: Principles 49
Principles of Garden Designing 49
1. Style 49
2. Balance 50
3. Focal Point 50
4. Space 51
5. Axis 51
6. Unity 51
7. Repetition 51
8. Rhythm 52
9. Variety 52
Characteristics of Plant Material 52
10. Divisional Lines 53
11. Scale and Proportion 53
12. Colour and Tone 53
13. Texture 55
14. Mobility/Transition 55
Availability of Light at Different Timings and Seasonal Variation 56

5. Basic Elements of Landscape Gardening: Landform Plant Material and Water 65
I. Landform 65
Landform: Categories 68

Functional Uses of Landforms 72
Aesthetic Uses of Landform 75
II. Plant Material 75
Functional influence of Plant Material 76
Architectural Influence of Plant Material 76
Aesthetic Influence of Plant Material 77
Visual Characteristics of Plant Material 77
Plant Size 78
Plant Form 79
Plant Colour 80
Growth Pattern 80
Plant Textare 81
III. Water 81
General Properties of Water 82
Functional Influence of Water 82
Visual Influence of Water 83
Water Park 85
Water Gardens 85

6. Components of Landscape Gardening 111
Definition and Significance 111
Garden Walls and Fences 111
Gates 115
Trellis Screens 115
Raised Beds 115
Hedges and Edges 116
Arches, Pergolas and Overheads 119
Arches 119
Pergolas 119
Overheads/huts 120
Lawn 120
Site and Soil Type 121
Selection of Grass Species for Lawn Development 121
Terrace 123
Children's Play Area 123
Steps and Ramps 124
Pavements 125
Seating 127

Carpet Bedding 128
Flower Beds 128
Shrubbery 129
Borders 129
Topiary 131
Garden Drives and Paths 132
Garden lights 133
Garden Decoratives or Adornments 134
Sculptures and Statues 134
Fountains 134
Floral Clocks 135
Bird Baths 135
Ornamental Tubs, Urns, Vases and Plant Stands 135
Sundials 136
Japanese Lanterns and Stones 136
Pillars 137
Colour Plates 147

CHAPTER 1

Landscape Designing and Gardening: Significance and Scope

Definition, Significance and Importance

Landscape design and gardening is the technique of design application to beautify our surroundings, our homes, office buildings and everywhere in open space with plants and garden features to give the soothing effect on the body, mind and soul. Rapid urbanization, industrialization with population explosion has been constantly depleting our natural resources through deforestation, ripping of hills and exposing them to erosion, pollution of rivers, streams and lakes with little understanding of the natural environment which supports human life. It is the application of the art and techniques of Landscape Gardening that can spell magic to our surroundings enhancing aesthetic beauty besides, improving and conserving environment. Besides, landscape beautification and ecological balance provision, other beneficial influence of gardens in human life have been researched and studied. Psychologists have found that access to plants and green spaces provides a sense of rest and allows workers to be more productive. Flowers convey our inner feelings sympathy, romance or celebration (pride and joy) or contrition in the most beautiful and delicate manner. Flowers are also used to express religious feelings. Interaction with gardens and natural spaces offers a variety of mental, physical and social beneficial influence for humans, ranging from stress reduction, quicker healing and mitigation of Attention Deficit Disorder in children to decreasing crime and air pollution. Studies throughout the world have proven the power of green spaces to improve human health. Recent studies in the Netherlands and Japan show that people with easy access to green space showed better health and lower mortality rates.

MABEL G. AUSTIN has rightly whispered about Nature in a poetry form,

"There is no greater loss in life to man,
Than being unaware at early dawn
Of Earth's awakening from a silver mist
Shot through with golden threads of breaking morn.
There is no greater sorrow in the world,
Than eyes unseeing, color everywhere,
Or ears unhearing, softly wafted notes
From Nature's great cathedral of the air.
There is no soul so dead as one of these,
Whose voyage leads through empty life, where hearts
Are veiled in darkness, claiming not the treasures,
Which Nature's beauty to the world imparts."

Gardening and Landscaping

Landscape is any piece of land on which it is possible to mould a view or design. Landscape Gardening is the application of garden forms, different styles, methods and materials with a view to improving the landscape. Landscape gardening has presently gaining great importance and developing with the growing awareness for environmental moderation along the developing concept of bioaesthetic planning. Thus, it deals with the conservation and understanding of the environment, with special relevance regarding the relationship between people and the natural environment, serving as a bridge between man-made and natural elements of the environment. Landscape architecture is the technique of designing gardens either with the help of pen or computer along with the structural components. Thus, it associates planning of building and landscape with great emphasis on scale, hard components, drainage and masonry work. Landscape architecture is thus, defined as the art and science of building on and with the earth's surface or more precisely it is the technique of planning, designing and portraying the ideas using different natural elements with the objective of improving and beautifying the functioning, architectural and aesthetic view of a landscape.

Importance and Benefits

The prime importance of landscape gardening lies in aesthetic developments and modernization of cities, town, villages, roadways, airports, railway stations, railway lines, bus terminus, city parks, and educational institutions against industrial fast growing pollution. Landscape gardening increases the awareness towards nature developed eco-friendly concept and upsurge the feeling of responsibility towards plants, birds and animals. It also helps in soil-moisture retention; prevent erosion, modifying air temperature, creating microclimate and removal of noise and dust pollution. It also provide habitat for birds and animals. A landform devoid of vegetation show erosive rate of around 500 tons/ m^2 /year. A green belt planting of around 183 m can reduce airborne particulate matter by as much as 75 percent. An acre of green area reduces 12-27 tones of dust, 1.5 tons of CO_2 and releases 2.5 tons of O_2. One fully developed large tree in a landscape can change the microclimate and brings down the temperature by 2 to 5°C. Parks, garden-spaces, street trees, and landscaped traffic islands provide more than a pretty panorama, effectively reducing the stress of our daily lives by invoking a feeling of tranquillity.The average temperature of concrete urban area devoid of green belt is higher compared to a thickly vegetated area. Well planned landscape around building or residence not only adjoins aesthetic value to it but also affix about 30% high value of real estate. Interaction with gardens and natural spaces offers a variety of mental, physical and social benefits for humans, ranging from stress reduction, quicker healing and mitigation of attention deficit disorder in children to decreasing crime and air pollution. Frances Kuo of the University of Illinois rightly quotes "Parks help people take care of themselves" advocating parks in cities represent a potential for a minor public investment with a huge payoff. She conducted a study of 28 identical high-rise public housing projects. She found that people living near green spaces: Boasted a stronger sense of community, coped better with everyday stress and hardship, were less aggressive and less violent, performed better on tests of concentration, managed problems more effectively. One national study of 450 children with Attention-Deficit/Hyperactivity Disorder determined that exposure to natural environments alleviated symptoms of the condition. Besides, environmental benefits, there are benefits of Landscape gardening on human psychology and health as well as Society and culture, children development etc.

Human Psychology and Health

Pleasant view of greenery and flowers stimulate creativity and has soothing effect on body, mind and soul. Landscape gardening is a healthy hobby, source of exercise for the body and eliminates any kind of mental problems or depressions. Studies have shown that stressed individuals feel better after exposure to natural scenes. Green spaces also reduce instances of aggression and violence. People feel the urge to view and enjoy the scenic beauty of green nature after being working in stressed urban concrete surroundings. Scientists have also reported that green spaces increase our ability to concentrate, both on the tasks at hand

and on our subconsciously-viewed surroundings. Doxis M. Palmer, a poet has depicted the feelings of garden Sanctuary in a verse as illustrated below:

"You who walk,

Maybe with troubled thoughts,

Come, enter here and rest;

And may the sweet serenity of growing things,

And the heavenly, peace

Be mirrored in thy soul."

Gardening promotes good exercise to the people engaged with it. Gardens offer a desired destination that inspire people for walking, meditation and exercising. For, health problems associated with Obesity and diabetes, mild and regular exercise like walking, *pranayaam* and some *yoga-asans* are recommended. Even relatively inert contact with nature—such as viewing it from a window—lowers blood pressure and anxiety levels. Besides, plants are source of nutrition like some berries and fruit trees as well as herbal medicines for human use like *Terminalia bellerica, T. chebula*, Neem, *Ginko biloba*, Moringa, *Madhuka indica, Murraya koeingi* , *Anthocephalus cadamba*, etc.

Society and Culture

Maintained green scapes play important role in improving social culture by increasing workplace productivity, work performance and creativity, safety (for driving, playing, get -together, privacy), reducing crime and prosperity. In the business environment, green spaces improve productivity and morale among workers. Studies show that desk workers with a view of nature—either out a window, or around them in the form of indoor plants feel more relaxed overall, and those with no visibility of plants suffer the most stress and anxiety. A study of 98 vegetated and un-vegetated apartment buildings in Chicago, US (2009) showed that vegetated spaces reduce crime by half, in addition to inspiring pride for surroundings that translated into less litter and less graffiti. Vegetated landscapes invite more people to use them, ensuring more eyes on the watch to prevent crime in outdoor spaces. Roadside vegetation or avenues also serve a social benefit by reducing fatigue, anger, aggression, frustration and stress of automobile drivers. Well designed landscapes and streets and shopping areas promote higher business activating as people love to explore and do shopping and raise opinions about the quality of goods and services offered. Further, face value and economic value of any building in well maintained landscape area may be 10-25% higher than the one without it.

Gardens and Children

Plants and nature play important role in child development. The book *Last Child in the Woods*: Saving our Children from Nature Deficit Disorder authored by Richard Louv (2009) has made an attempt to bring together a new and growing body of research indicating that direct exposure to nature is essential for healthy childhood development and for the physical and emotional health of children and adults. More than just raising an alarm, Louv offers practical solutions and simple ways to heal the broken bond with nature right in our own backyard. Richard Louv (2009) in his book, has truly reviewed the benefits plants offer to the growing child. Views of trees from the home improves self-discipline among inner city girls, including enhanced concentration, inhibition of impulsive behavior and delay of gratification. There is clear difference in the mindset, behavior and physique of a child playing in a green surroundings and the one who is stuck on TV or computer screen for hours, where the former is energetic, quiet and physically slim and fit while the other is impulsive, lazy and heavier in weight. After creative play in verdant settings, children overall demonstrate increased ability to concentrate, complete tasks and follow directions. Besides, such areas will encourage kids for creative ideas like drawing, painting or playing music. Kids acquire the sense of responsibility and develop managerial skill by playing and taking care of his/her own garden along with acquiring knowledge of plant identification, growth habit, uses and handing of fresh produce. Research now suggests that thoughtful exposure of youngones to nature can even be a powerful form of therapy for attention-deficit disorders and other maladies. We can rightly assume as assured by the scientists that just as children need good nutrition and adequate sleep, they may very well need contact with nature. Gardening inculcates the bonding between the nature and children during the early stages of life that is reflected in his/her character building in later life.

Other Benefits

Other benefits of landscape gardening includes

1. *Creates indoor or outdoor space as well as defines space* for beautification and architectural effects
2. *Promote serenity and spiritual well-being*: For many people, being in nature and interacting with the natural world, brings a sense of peace, tranquility and feeling of being integral part of the universe and a higher power.
3. *Encourage social interaction*: Gardens form the platform for interaction among people irrespective of any age, class, caste or colour. Gardens can encourage this interaction if they are easily accessible to patients, families, and staff and offer groupings of lightweight, moveable chairs.

4. *Gardens enhance a sense of Space* :Gardens can enhance a sense of control if they offer a variety of spaces to choose from-some private and some open, some sunny, some shady, some with background sounds, some without, and so forth among people.

Besides this, gardening can also enhance self-esteem, because plants respond to care given by people. Promotes cleanliness, organized culture and calmness in the society and promote creative activities like nature photography, nature painting and bird watching as gardens are habitat for birds.

Professional Scope

The landscaping profession conveys and directs to make human life more elegant, satisfying, eco-friendly and productive. Landscape gardening employs thousands of people through various activities like nursery and rental plant services, besides landscape designing. Landscape gardening also plays a vital role in giving emphasis to the tourist business, now known as ecotourism by improving aesthetic and functional uses of historical and religious places, sea beaches, rivers and dams, hill stations and other tourist places. The profession of landscape architecture is a design discipline that combines art and science with primary focus on merging people and their activities with land. The wide areas where one can develop his/her career in landscape gardening with the skill of green fingers are as given below:

1. Horticulturist
2. Landscape developer and manager
3. Landscape consultant
4. Landscape designer
5. Landscape contractor
6. Nursery Owner/ man
7. Plant rental service
8. Bedding plant industry
9. Gardener
10. Tourism

Plate 1: Landscape Gardening: Importance and Benefits

Plate 2: Landscape Gardening: Importance and Benefits

CHAPTER 2

Basic Garden Styles and Conventional Gardens

Gardens have been considered important, sacred from ancient time and the styles developed being inspired form its surroundings. Thus the influences of climate, general landform, religion, spiritual vision, culture, nation, prevailing political conditions, social factors, architectural styles, etc have influenced in developing specific garden style in a particular region. These styles developed slowly over hundreds of years and remained insulated and remained intact till the travelling approach between different countries got well established. Later as the mobility between different countries became widespread, these style forms became intermingled and refined even to new forms. This resulted into the development of Japanese garden style or the arts of 'bonsai' making which had its origin in China, or the creation of west coast American garden that originated from the Bauhaus and modern movement that stated life in Germany, the mughal garden style in India based on Persian style from Iran etc.

Basically, the different garden styles can be classified into formal, informal, intermediate and wild garden styles. Aristotle, the father of mathematics and Pythagorus, the father of geometry and disciple of Aristotle, strongly influenced the geometry, mathematics and calculative features of formal garden style. While informal garden style have evolved from the eastern vision of thoughts, originated with Laotsu and Confucious contemplation (preaching) that manifested gardens to express peaceful and compassionate atmosphere influencing meditation, silence and internal harmony.

Formal Garden Style

A formal garden is laid out in a symmetrical or a geometrical pattern. In this garden the design is stiff as everything is done in a straight and narrow way. In such gardens everything is planted in straight lines. If there is a plant or flower bed on the left hand side of a straight road, a similar plant or flower bed must be planted at the opposite place on the right hand side. This style includes Mughal gardens, Persian gardens, Italian gardens and French gardens.

Key features of formal style: i)Plan is made on paper and land is selected accordingly. ii) The layout is symmetric with square/ rectangular shape and roads cut at right angle. iii) It has some sort of enclosure or boundary. iv)Flower beds also have geometric designs. v) The arrangements of trees and shrubs are also in geometry and kept in shape by regular pruning and training. vi) Features like fountains, pools, cascades, etc. are used for attraction.

Egyptian Gardens

Formal gardens existed in Egypt as early as 2800 BC. These were also unique in their time as they were developed in dry-arid climate of Egypt, representing technique and skill developed during that time. Egyptian tomb paintings of the 1500s BC are some of the earliest physical evidence of ornamental horticulture and landscape design. Lotus ponds surrounded by symmetrical rows of acacias and palms are clearly depicted in those paintings. Egyptian Gardens (2000-1200 BC) were treasured, considered sacred developed near the temples. Herbal and vegetable plots, vineyards, flower beds and ponds were common features of the temple gardens. Tree groves and avenues leading to the temple entrance were also unique feature. During the eighteenth dynasty of Egypt, gardening techniques were fully developed and used to beautify even the private residence. Private home gardens with fruit trees, vegetables, flowers, grape vineyards and fish ponds were cultivated by people depending on their financial status.

Persian Gardens

Persian gardens represent the Islamic and Judeo-Christian ideals of paradise on earth and were believed to symbolize Eden and the four Zoroastrian elements of sky, earth, water and plants. The main feature of a Persian garden design was *nahars* (flowing canals) of water – the concept of Persian Paradise, "where cooling water flows". The geometrical design of the Persian garden, and its survival in extreme dry and scorchy climatic conditions, reflect the application of technology, water management and engineering, architecture, botany and agriculture during that time (800-600 BC). A wide variety of species, water channels, fountains and pools all in geometrical shapes, mixed with soft musical knots, enhanced with elegant architectural competence were important features that not only fulfilled the urge of pleasure and fun but were a source of inspiration for creative activities be it poetry, contemplation and seclusion. The notion of the Persian Garden permeates Iranian life and its artistic expressions in literature, poetry, music, calligraphy and carpet design. Water as representation of Self reflection, Meditation, Sacredness and tranquility was an important element of these gardens. Other features included buildings, pavilions and walls, as well as sophisticated irrigation systems. They have influenced the art of garden design as far as India and Spain. The perfect design of the Persian Garden, along with its ability to respond to extreme climatic conditions, is the original result of an inspired and intelligent application of different fields of knowledge, i.e. technology, water management and engineering, architecture, botany and agriculture. The main principles of persian gardens were (i) Fusion of human creativity and intelligence reflected in the perfection of the design of the Persian Garden, based on the right angle and geometrical proportions that was often divided into four sections known as *Chahar Bagh* (Four Gardens) with sophisticated water-management system, as well as the appropriate choice of flora and its location in the garden layout in the desert setting. The *Chahar Bagh* was a reflection of the mythical perception of nature. (ii) These Gardens exhibit an important interchange of human values, having been the principal reference for the development of garden design in Western Asia, Arab countries, and even Europe. (iii) These gardens became a central feature in private residences, palaces and public buildings, as well as in ensembles associated with benevolent or religious institutions, such as tombs, park layouts, palace gardens, Meidans, etc. Reflection of social and cultural traditions that have evolved in Iran and the Middle East over some two and a half millennia was depicted in these gardens. (iv) The Persian Garden is an outstanding example of a type of garden design achieved by utilising natural and human elements and integrating significant achievements of Persian culture into a physical and symbolic-artistic expression in harmony with nature. (v) The Persian Gardens can be directly associated with cultural developments as source of inspiration for higher values like poetry, miniature painting, music, architectural. These include literary works and poetry by Sa'di, Hafez and Ferdowsi. The Persian Garden influenced the Persian carpet and textile design, architecture, ornaments, etc. In the *Avesta*, the ancient holy book of the Zoroastrians, the Persian Garden and its sacred plants are praised as one of the four natural elements (earth, heaven, water, and plants). Some examples of the Persian gardens are Pasargad Persian Garden at Pasargadae, Chehel Sotoun, Isfahan, Fin Garden, Kashan, Dolatabad Garden, Yazd in Iran, Humayun's Tomb, New Delhi, India.

Important elements of Persian garden revolved around modifying the effects of sunlight and through shade (trees and trellises largely feature as biotic shade; pavilions and walls) and water (in form of *qanat*, well, planting of trees in a ditch i.e *jub*, etc). The Persian style also attempted to integrate indoors with outdoors through the connection of a surrounding garden with an inner courtyard with clear arches.

Styles: The primary styles of the Persian gardens are a) *Hayât*- Refers to 'Living in present', is a classical Persian layout with heavy emphasis on aesthetics over function. In Public gardens, man-made structures in the garden are particularly important, with arches and pools (which may be used to bathe). The ground is often covered in gravel flagged with stone. Plantings are typically very simple - such as a line of trees, which also provide shade. Privately, these gardens are often pool-centred and, again, structural. The pool serves as a focus and source of humidity for the surrounding atmosphere. There are few plants, often due to the limited water available in urban areas. b) *Meidan*-This is a public, formal garden that puts more emphasis on the biotic element than the hayât and that minimises structure. Plants range from trees, to shrubs, to bedding plants, to grasses. Again, there are elements such as a pool and gravel pathways which divide the lawn. When structures are used, they are often built, as in the case of pavilions, to provide shade. Eg Naghsh-i Jahan square, the charbagh Royal Square (Maidan) in Isfahan, constructed between 1598 and 1629. c) *Chahar Bâgh*-These gardens are private and formal. The basic structure consists of four quadrants divided by waterways or pathways. Traditionally, the rich used such gardens in work-related functions (such as entertaining important guests and visitors). These gardens balance structure with greenery, with the plants often around the periphery of a pool and path based structure. d) *Park*-The Persian park serves a casual public function with emphasis on plant life with the purpose of relaxation and socialisation. They provide pathways and seating, but are otherwise usually limited in terms of structural elements. e) *Bâgh*- It emphasizes the natural and green aspect of the garden. It is a private area with the primary function of family relaxation. It is affixed to houses and consist of lawns, trees, and ground plants. The waterways and pathways stand out less than in the more formal counterparts and are largely functional.

Mughal Gardens

The Mughal gardens in India were laid out during the rule of Mughal Emperors in India, these were inspired by Persian garden style. Babar (1494-1531) was the first Mughal ruler to introduce this style in India. All other rulers and some of the Mughal Begums also developed these Mughal gardens. Famous Mughal gardens of India are Vrindhavan garden at Mysore, Ajwa garden at Vadodara, garden at Taj Mahal, Agra, Nishat garden, Kashmir etc. The main features of Mughal gardens, which are largely borrowed from the Persian style, are: (a) site and style of the design, (b) walls, (c) gates, (d) terraces, (e) *nahars* or running water, (f) baradari, (g) a tomb or a mosque, and (h) trees. Some important aspects of (a) Site and design: The Mughal were very choosy about the selection of site and always preferred a site on a hill slope with a perennial rivulet or along the bank of a river. Mughal gardens are generally rectangular or square in shape and different architectural features form the main base of the design.(b) Walls and Gates: The Mughals created the gardens not only for pleasure and recreation but for protection of forts or residence with high walls and with-an-imposing wooden gates. The purpose of the high walls was security from the enemies and shelter against hot winds. This also provided a sense of privacy in the gardens.(c) Terraces: Mughals were fond of terraces in the gardens because they came from the hilly terrains and hence they used to select the location of gardens near hill slopes. According to Islamic faith the Paradise has eight divisions so the gardens have eight terraces corresponding to the eight division or occasionally with seven terraces which represent the seven planets.(d) Nahar or Running Water: The style of having running water by constructing canals and tanks was borrowed from the Persian garden style and were paved with tiles or marbles of blue colour to create an illusion of depth. The course of water used to be moved in various ways taking advantage of each slope and further ending of flow was into artificial falls and ripples. The water canals were used to have fountains which were illuminated with small lamps to create beautiful reflections. In the plains of India, where the summers are hot, water was utilized for its cooling effect.(e) Baradari: This is nothing but an arbour-like structure, made up of stones and masonry with a *pacca* roof and a raised platform for sitting. These have twelve or occasionally more doors on all sides from where the emperors could see the view of whole garden as well as dance performance on several occasions. (f) Trees and Flowers: The trees were selected with careful planning as each tree symbolized the basic essence of life like fruit trees symbolized life and youth, while cypress represented death and eternity. The seasonal flower beds were of geometrical pattern and constructed along the water canals or near the main buildings. Plants having fragrance like rose, jasmine, etc. were preferred. Besides these, carnation, hollyhock, delphinium, chrysanthemum, etc. were also grown. (g) Tomb or Mosque: It was a common practice to have the gardens built around a tomb (e.g., Taj Mahal at Agra, Akbar's Tomb at Sikandra). Mughal gardens were at their paramount when built created a monument.

Greek/Hellenistic Gardens

The Greek gardens featured trees and shrubs around temples and highlighted paths and roads with statues but they did not develop private gardens for pleasure and fun as revealed from historical findings. The greek did not believe in training and maintaining practices for gardens but left them more natural to flourish as they were not man made things. However, the unique Hellenistic garden was that of the Ptolemaic dynasty (325–35 BC) in Alexandria, a grand enclosed walled beautiful landscape that included the famous Library, the largest and most significant great library of the ancient world and the famous museum that was home of music, art and astronomical observations. The Greek mathematician Heron developed water systems on this garden as revealed from ancient writings on hydraulics and mechanics .

Roman Gardens

Romans are known for growing plants for ornamental purposes and developed pleasure garden to showcase the landscape with many beautiful features and breathtaking views for the basic purpose for relaxation and rejuvenation. Most Roman villas featured a transition element between the house and the garden. The Romans incorporated marble into their patio type structures that opened to a grand landscape that was designed in all directions, pergola style overhangs decorated with vines and hanging flowers. A walk would be the main element of the grounds for seeing each garden delight. Generally, the property would feature a terraced section, an orchard or vineyard (probably both), a kitchen garden for herbs and vegetables (this might not be a part of the grand garden tour however) and various sections that might contain prized plants like roses. The center of the garden would also feature a special element like a fountain with statues. Other important features might include shrines and grottoes. Water was prized in all styles of Roman gardens. From the central fountain to a stream running down the terrace, to a garden pond to an elaborate pool that was an important social element of the Roman world. As for the type of plants typical Roman choices include roses, cypress, rosemary, mulberry and fig trees, dwarf variety trees, tall trees, marigolds, hyacinths, narcissi, violets, saffron, cassia, thyme and many more.

The Roman garden was an important area of the house and would often be used as an outside dining and entertaining area. Gardens were usually enclosed by walls or even surrounded by the house, creating a central courtyard. Sometimes colonnaded walkways would surround areas of the garden, including rectangular lawns. Romans used garden ornaments, such as statues (particularly of gods), urns and sundials. Pools and fountains were created in the garden, usually in a rectangle or a semi-circle shape. Pergolas, which were constructed to create shade, were an attractive feature with vines or roses climbing up them. Arbours and dovecotes also featured. *Aegopodium podagraria,* is believed to have been introduced to Britain by the Romans as a salad crop. Centuries later Roman gardens were to have a huge influence on the Italian renaissance period in the 15th and 16th century. Roman gardens besides being a place of peace and serenity depicted religious and symbolic meanings. Technical knowhow on horticulture, animal husbandry, hydraulics, and botany was actively exchanged among people and planting material and plants were widely shared. The Gardens of Lucullus (Horti Lucullani) on the Pincian Hill on the edge of Rome introduced the Persian garden to Europe, around 60 BC. Private Roman gardens were generally separated into three parts, *viz.*, a) the *xystus*, a terrace that served as an open air drawing room and connected to the home via a covered portico. Second , b) *ambulation* (lower garden) with a variety of flowers, trees, and other foliage and served as an ideal for recreation activities and c) the *gestation* , a shaded avenue (surrounding the ambulation) for walking or horse riding. Excavations in *Pompeii* show that garden size was reduced so that even a layman can have enjoy it in the surroundings of his average home. Even tall buildings (*insula*) were adorned with *window boxes* or *roof gardens*. Modified versions of Roman garden designs were adopted in Roman settlements in *Africa, Gaul*, and Britannia. Roman garden designs were later adopted by Renaissance, *Baroque*, *Neoclassical*, and even 20th century landscape architects.

The Italian Renaissance Gardens

The Italian Renaissance garden was a new style of garden which emerged in the late 15th century at villas in Rome and Florence, inspired by classical ideals of order and beauty, and intended for pleasure. Italian gardens are known for their grandiose yet tranquil ambience. In the late *Renaissance*, the gardens became larger, grander and more symmetrical, and were filled with fountains, statues, *topiaries*, water bodies and other features designed to delight their owners and amuse and impress visitors. Italian garden design adopted the use of art topiary to establish strong, evergreen structure in the Italian garden. The style was imitated throughout Europe, influencing the gardens of the French *Renaissance* and the English garden. There were 1,168 different plants and trees, including a fan palm tree brought from Egypt. In 1545, in Florence, *Giardino dei Semplici*, the garden of medicinal herbs was developed. Soon the medical schools of the universities of *Bologna*, *Ferrara* and *Sassari* all had their own botanical gardens filled with exotic plants from around the world. Villa dEste, Villad Tolle, Villa d'Costello, Sarco Bosco, Villa Lente were the famous Italian garden styles. Due to the difference in climate from the cold and wet north to the warmer and almost arid south, the gardens in various parts of Italy vary in their vegetation and structure. This has led to the popularity of fusion Italian gardens in modern times, namely the Tuscan gardens. Some features of Italian garden include a shaded patio area near the house where one can sit and dine with easy access to the house and all its amenities, simple pergola, with vines, roses and scented plants like *Rhynchospermum jasminoides* grown over it, with Italian-style furnishing like rusted iron chandeliers and candle holders. Tough, plants plants like Oleander (Nerium oleander), *Viburnum tinus* and

Pittosporum tobira, Cupressus sempervirens, white Hydrangeas and elegant fragrant evergreens like *Osmanthus fragrans* and *Gardenia jasminoides* were favourite in Italian gardens. Plants like Boxwood (*Buxus sempervirens*), Yew (*Taxus baccata*), Holm oak (*Quercus ilex*) and Bay laurel (*Laurus nobilis*) were exclusively selected for topiary work.

The French Renaissance Gardens

A French garden is a very specific style of formal garden. The designers of the French gardens used architectural aspects, with strict rules of geometry, optics and perspective. Gardens were often served as theatres and designed like green buildings i.e more of architectural aspects used and displayed through plants. French gardens are used most classically in the landscaping of large formal structures like museums, private mansions, and so forth, although one could also install such a garden in front of a more modest structure. Symmetry and order are very highly valued, with all hedges, lawns, trees, plants, and shrubs being meticulously maintained. These gardens were characterized by symmetrical and geometric planting beds or parterres; plants in pots; paths of gravel and sand; terraces; stairways and ramps; moving water in the form of canals, cascades and monumental fountains, and extensive use of artificial caves, labyrinths (Jumble network) and statues of mythological figures. The distinguishing feature of a French garden is that it is centered on the facade of a building, differentiating it from many other formal gardening styles. The French garden draws the eye to the building, and integrates the building into the landscape with its very geometric style. These gardens also typically have numerous reflecting pools, fountains, and ponds, with gravel and lawn paths to allow people to navigate the garden. Clipped evergreens are a common feature in French gardens, bordering paths and flower beds. Many French gardens are also dotted with small ornamental buildings, which may range from open-air gazebos for summer entertaining to fully enclosed structures. Historically, each structure had a very specific purpose, with these buildings being used to play board games, serve tea, listen to music, paint, and engage in other recreations in the garden. Such buildings were also used by royalty for formal audiences with guests. Maintaining a traditional French garden needs constant trimming, pruning, weeding, and other maintenance tasks, as it must look immaculate at all times. New flowers are constantly being planted to replace worn and tired plants, and the design of the garden is constantly being refined with small and subtle details.

The French Classical garden style, or *Garden à la française,* that was grander and more formal climaxed during the reign of *Louis* XIV of France (1638–1715) and his head gardener of *Gardens of Versailles, André Le Nôtre* (1613–1700). The french gardens epitomize monarch and 'man' domination even on manipulating nature.

Informal Garden Style

In an informal garden style, the whole design looks informal, as the plants and the features are arranged in a natural way without following any strict rules. But here also the work has to proceed according to a set and well-thought-out plan otherwise the creation will not be artistic and attractive. The idea behind this design is to imitate nature. Chinese gardens and Japanese gardens are based on this style.

Key features of informal style: This style is more natural and holistic with soft appeal, it is in contrast to the formal design, plan is asymmetric according to the land availability for developing the garden, smooth curvaceous outlines are applied extensively in different forms and water bodies and flower beds are irregular in shape.

Chinese Gardens

The Chinese gardens, also known as a Chinese classical gardens followed a style that deeply depicted nature same as it has inspired other forms of chinese art. The art of garden design began in the Shang Dynasty (1600 BC – 1046 BC) as large imperial parks. Chinese gardens were meant to connect with nature and an idealistic way of life. Plants were used as symbols *viz.*, Bamboo -strong but resilient character, pine - longevity, persistence, tenacity and dignity, The lotus symbolized purity, plum blossom -renewal and strength of will, chrysanthemum - symbolize splendor and luster Peonies – wealth. Banana trees were used simply for the sound they make in the breeze and flowering peaches were grown for spring color, and sweet olive as well. Chinese garden's structure is based upon the culture's creation myth, rooted in rocks and water. Thus, Chinese landscape is known as *Shan* (mountain) and *Shui* (water). The essential elements of chinese garden are a wall surrounding a hall, a pool, and trees and mountains. Later these were expanded to seventeen essential elements: 1) proximity to the home; 2) small; 3) walled; 4) small individual sections; 5) asymmetrical; 6) various types of spatial connections; 7) architecture; 8) rocks; 9) water; 10) trees; 11) plants; 12) sculpture; 13) borrowed scenery; 14) chimes; 15) incense burners; 16) inscriptions; 17) use of *feng shui* for choosing site.

Japanese Gardens

Japanese gardens finding their close association with nature, were developed under the influences of the distinctive and stylized Chinese gardens. One of the great interest for the historical development of the Japanese garden, bonsai and related arts is Zen monk *Kokan Shirena* (1300 BC) in his essay *Rhymeprose on a Miniature Landscape Garden*. In Japanese culture, garden-making is a high art, intimately related to the linked arts of calligraphy and ink painting. During the Asuka period (538–710), gardens were supposed to express Buddhism and Taoism through replicating the mountainous regions in China (Japanese Lifestyle). Ruins of these types of gardens can be found in Fujiwara and Heijyo castle towns (Japanese Lifestyle). During the Heian period (794-1185), gardens shifted from solely representing religious beliefs to becoming, "a place for ceremonies, amusement, and contemplation". In the Kamakura and Muromachi periods (1185-1573), a great many gardens were created during these two time periods due to improved garden techniques and the development of Syoinzukuri style (Japanese Lifestyle). Zen beliefs were also flourishing at this time and had great influences over garden techniques and purposes. A notable gardener who appeared during these periods is Soseki Muso: He made Saihoji Temple (Kyoto), Tenruji Temple (Kyoto), and Zuizenji Temple (Kamakura) gardens. Sen no Rikyu (1517-1591) created the traditional style of a tea house where there was usually a roji ("dewy path") leading to the house (Japanese Lifestyle). Besides the tea houses, gardens constructed in the Edo period (1603-1868) reflected the tastes and style of each individual shogun ruler. These tea house styled houses and gardens can be seen in Kenrokuen (Kanazawa), Kôraku-en (Okayama), Ritsurin Garden (Takamastsu), Koishikawa Korakuen (Tokyo), and Suizenji Park (Kumamoto) in Japan.

The tradition of Japanese gardening was historically passed down from *sensei* (master) to apprentice. It was a highly technical art and only those well versed and educated were allowed to design gardens. In recent decades this has been supplemented by various trade schools. Japanese gardens often contain several of these elements like water, real or symbolic (in form of sand or pebbles arranged like ripples), a bridge over the water body, or stepping stones, rocks or stone arrangements (or settings), a lantern, typically of stone, a teahouse or pavilion, an enclosure device such as a hedge, fence, or wall of traditional character. Typical Japanese gardens have at their center a home from which the garden is viewed. In addition to residential architecture, depending on the archetype, they are further classified on the basis of positions, shape and purpose. The important types are:

i) ***Hill Garden*:** This style in Japanese is known as *Tsukiyama-niwa* or *Tsukiyama-sansui*, meaning hills and water. The hill garden is made up of one or more hills designed with earth mounds and exposed weathered stones. Water is used in the form of a stream or a pond or waterfalls or all the three with or without islands and also bridges, lanterns, stones and trees. The important points in the garden are that they are decorated with stones and selected trees. Untrimmed stepping- stones are placed over the paths.When the island is present it should be decorated with a "Worshipping stone", called *raithai-seki*, a "Snow-viewing" lanterns and a pine trees.

ii) ***Flat Garden:*** These are laid out in flat ground without hills or ponds which is also called as *Hira-niwa*. Flat gardens are supposed to represent a mountain valley or a meadowland. There should not be any ups and downs in a flat garden though the low rounded hills can be designed with the help of stones or earth mounds or both.The usual features are a well, a water-basin made of stones, stones lying close to the ground, stepping stones and trees. The trees are trained to lie close to the ground. In a flat garden, the principle is to avoid strong vertical lines represented by tall pines.

iii) ***Tea Garden:*** The tea garden is laid out based on principle of the Japanese tea ceremony and hence needs space of at least about 200 square meters for its designing. It is essential that the garden be enclosed by a fence to create an atmosphere of intimacy with a gate made of very light material such as bamboo. The tea gardens are further divided into an outer garden (*Soto-roji*) and inner garden (*Uchi-roji*) to protect it from the noise of the outer world.The outer garden is a waiting place where the guests are supposed to wait until the host appears to welcome them. While the inner garden contains the tea house, a small straw hut. The entire path to the tea house is paved with stones or studded with stepping-stones. The most important feature at the entrance of the tea house is water –basin or a well or both for the visitors to rinse their face before entering for the ceremony. To illuminate the water basin and resting place, stone lanterns are set at appropriate places. The outer garden will have simple plantings of deciduous trees and stone groupings with water basins. The inner garden has evergreen trees casting more shadows. The entrance to the tea house is through a low-door so that the guests have to enter in a bending posture, simulating respect and humanity.

iv) ***Passage Garden:*** The passage gardens, the *Roji-niwa*, are those which are laid in narrow passage as for example a narrow space between two houses or approaches to buildings. The common features of a passage garden are a few key rocks, slabs of stones, and only a couple of type of plants. Plants with open form and slender shapes are selected. There should be hardly any ornaments such as lanterns, basins or other man-made feature.

v) ***Sand Garden:*** It is the simplest style of gardening. The most famous sand garden exists in Kyoto and is known as *Ryoanji* garden. The garden consists of a rectangular area of about 350 square meters. The main feature of this style is to arrange a few vertical and prostrate stones in groups of 2 or 3 and to fill in the gap between the stones with fine white gravel. The ripples of flowing water can be made with rake. The raking has to be repeated often to keep the garden in its best shape. The important features of Japanese Gardens are given:

a) *Ponds:* Ponds are of irregular shape. The banks are generally bordered with stone piling work in a regular or an irregular fashion. They are made by peddling clay, or using concrete as a bottom. The ponds are generally fed by a stream or a waterfall.

b) *Streams:* Small streams are arranged most naturally with natural stones bordering the banks. Fanciful stones are arranged in midstream to break the flow of water or by changing the inclination.

c) *Waterfalls:* A waterfall may be made effective by manipulating it to drop in two or three levels. To make it more natural, large-sized stones are arranged around waterfalls. These are partially screened by planting a group of trees in front of the waterfall. The dense planting of evergreens around the waterfall also symbolizes a mountain scenery.

d) *Fountains:* Often natural fountains are provided near the foot of the hill, on the hill-side or in the forest. Sometime water is conveyed from a hillside by means of bamboo-piping.

e) *Wells:* This is the feature of utility and for adornment. They may be square, circular, or criss-cross in shape. The frame is generally constructed of stone or wood provided with a pulley is also used. The buckets are hung on either side, suspended from a rope. The well is also very ornamental.

f) *Island:* There are four important garden islands, the first two types representing sea islands. First is "Elysian Isle" (*Horai-jima*) which is constructed in the middle of a lake and is never connected by a bridge. The beach is decorated with sand, shells and pebbles. Often this island is given the shape of a tortoise. The second type of island, the "Wind-swept Isle" (*Fukiage-jima*) and constructed in a similar way. The beaches are decorated with sea rocks, sand and shell. The other two islands are "Master's Isle" (*Shujlin-to*) and the "Guest's Isle" (*Kiakujin-to*). The former is placed in the foreground of the landscape so that it can be easily approached by a bridge from the bank. The latter is located in the background and is accessible by bridges and stepping-stones. The "Master's Isle" generally has a summer house which is a thatched arbour with some selected trees, stones, stepping-stones and one or two lanterns. Often the islands are in the shape of mounds or hillocks. A few other types of island are "*Mountain Isle*", "*Forest Isle*" and "*Rock Island*".

g) *Bridges:* Bridges are a special feature in a Japanese garden. Bridges may be made of stones, polished or unworked, earth, wood, and other materials but the construction should be pleasing to the eye. The aim is to prolong the crossing time so that the visitor gets enough time to enjoy the scenery round. A combination of stepping-stones and bridges is also often used to reach an island or to cross a stream. If there are two or more bridges in the garden, different types should be constructed to bring variety. When the pond or waterway is large enough to permit boating, the bridges should be of arching type to permit the boats to pass under.

h) *Water-Basins:* The water-basins are fitted near a house meant for the guests to rinse their mouth and wash the hands. But now a day these are used as ornamental features. The distance and size of basin are in proportion to the size of the building. A small house may have a basin 1 m tall, whereas in front of a large house the basin becomes as tall as 2 to 2.5 m and it is useless but remains as an ornamental feature. The water-basin comes in various shapes, the most common ones are in the shape of an urn, square star shaped, cylindrical, stone-bottle shaped, and bowl shaped. A screen-fence is provided near the water-basin to screen off unwelcome sight. Stones are placed at the base to arrest the splash of water. A lantern is provided near by for illumination.

i) *Stone lanterns:* It is an important feature of any Japanese garden. The usual stone is granite, but sand stone or white marble also known as "*Snow-Scene*" may also be used. The usual places of fixing the lanterns are near the base of a hill, on an island, on the banks of a lake, near a water-basin or a well, along a path, on a boat-landing, near waterfalls and a bridge. The lanterns are used singly but along with a combination of rocks, fences, water-basins, shrub and trees.

The lanterns are not exactly meant for illumination but as objects of ornaments. The object of illumination then becomes a dim mysterious glow. A lantern has six parts, namely, the ornamental top, cap, light chamber, middle stand, post and base.

j) *Stones:* Stones are selected according to size shape, and colour. The stones are rarely in isolation but rather arranged in groups of two to five. All stones must be arranged with a firm foundation as stones of unstable nature show the weakness of a garden design. The principle is to make them look natural and for this purpose, low-growing bushes or upright trees are planted near the stone grouping. Naturally, the size, shape and colour of the stones vary according to the purpose and the place of their use.

k) *Pagodas:* Another favourite feature of Japanese landscape is the stone tower or the pagoda which is a structure consisting of two, three, five or more separately roofed stages.

l) *Fences and Gates:* Fences are of two types, one is for partition which should look light in appearance and hence wood and twigs of bamboo are preferred to stone, while the other is for the purpose of screening which is called "*Sleeve Fences*", as its shape resembles that of the sleeve of a lady's Kimono. There are generally two gates, one is the front entrance and second the back entrance. Gates are also made of light materials such as wood or bamboo. Some gates are bare while others are roofed. The roof may be made of bamboo, wood, or simply thatched.

m) *Vegetation:* After the stones, the secondary garden framework, i.e., the evergreen plants are arranged. The trees are of permanent nature and stand as reference points in the garden. The aim of a Japanese garden is to imitate nature by using natural elements, and hence, there is hardly any bar in using any plant material.

Intermediate Garden Style

An intermediate garden style is based on the principles of formal and informal gardens. Many Indian gardens are of intermediate style. The main features of this type of gardens are rockeries, borders, lawn, shrubbery, winding walk, group of trees, waterfalls, etc. Some features may follow strict geometrical outline but it is not bounded for all features. This includes English garden which is one of the popular intermediate garden style. This garden style is not only for enjoyment but also for utility like growing of fruits and vegetables for home requirement. This style of garden provides an opportunity to designer to use his own concept by referring formal and informal style. This type of garden is generally seen in small towns where people themselves develop their own garden style with their ideas. English gardens are categorised in the intermediate style of garden.

English Gardens

The English garden presented an idealized view of nature, inspired by paintings of landscapes by Claude Lorraine and Nicolas Poussin, and the classic Chinese gardens of the East, which had been described by European travelers. English gardens are based on informal countryside look. English gardens are one of the most beautiful European gardens. The English gardens are extremely fine, and are formed with an agreeable wildness and pleasing irregularity. There used to be no barrier of hedge or fencing in old times. The main features included a lake, lawns set against groves of trees, and recreations of classical temples, Gothic ruins, bridges, and other picturesque architecture, designed to recreate an idyllic pastoral landscape. Kitchen gardening with some herbs growing were also preferred. Later on the concept of flower beds, topiary and terrace gardening also developed. Today, the main features of English gardens are lawn, rockeries, herbaceous border, rock gardens, mixed annual borders and shrubbery. Water rivulets and fountains, curved paths, informal group of trees and clipped hedges have also gained importance. The novelty and exoticism of Chinese art and architecture in Europe led with works of William Chambers (1723–1796) in form of a book about Chinese arts (1757). In 1761 he built a Chinese pagoda, house and garden in Kew, London, as part of Kew Gardens, a park with gardens and architecture symbolizing all parts of the world and all architectural styles and new anglo-chinese style evolved. By the end of the 18th century the English garden was being imitated by the French and the Russians as well. The most influential figure in the later development of the English landscape garden was Lancelot Brown also known as Capability Brown (1716–1783) who began his career in 1740 as a gardener at Stowe under Charles Bridgeman, then succeeded William Kent in 1748. The gardens were simplified by eliminating geometric structures, alleys, and vacant space near the house and replacing them with rolling lawns and extensive views out to isolated groups of trees, making the landscape seem even larger. Artificial lakes were created and used dams and canals to transform streams or springs into the illusion that a river flowed through the garden. Brown designed 170 gardens, of which the most famous were *Petworth* (West Sussex) in 1752; *Chatsworth* (Derbyshire) in 1761; *Bowood* (Wiltshire) in 1763 and *Blenheim Palace* (Oxford shire) in 1764.

The 'Gardenesque' style of English garden design evolved during the 1820s where the positing of plants for optimum display was emphasized and small-scale landscapes, with features and adornments, to enhance beauty as whole, add variety and to create mystery. New species from other countries were introduced. The Gardenesque approach involved the artificial mounds helped to stage groupings of shrubs, and island beds became prominent features."Wild" gardens and herbaceous borders became popular in the mid to late 19th century as described in the books of William Robinson and the pictures by Kate Greenaway. In the last quarter of the 20th century, less structured Wildlife gardening emphasized the ecological framework of similar gardens using native plants. Spring in English gardens are well known for their intoxicating beauty. Flowering annuals were the charm of English gardens. Bulbous plants like Narcissus, Daffodils, Iris, shrubs like Magnolias, Rododendrons, fuschia, Hydraengea, Roses, herbaceous like Peonia, Daisy, Climbers like *Wisteria multijuga*, etc. Annuals like Petunia, Phlox, Hollyhock, Zinnia, Poppy, Marigold, Snapdragon, Nasturtium, Larkspur, Foxglove, Forget Me Not, etc were common in English gardens.

Wild Garden Style

Wild Garden is comparatively recent style of gardening, which developed in the last decade of the nineteenth century. The concept of wild garden is not only against all formalism but it also breaks the rule of landscape styles. The main idea was to naturalize plants in shrubberies. The grass should remain unmoved, as in nature, and few bulbous plants should be grown scattered in the grass to imitate wild scenery. The passages should be opened in the woodland, and trees, shrubs and bulbous plants should be planted among the forest flora to fulfill the idea of a wild garden. The creepers are allowed to grow over the trees naturally imitating those of the forests.

The plants used for woodland of wild garden should be shade and water stress tolerant, hardy and should not require tying or staking. Generally, the plants of wild garden are local species for easy establishment. Trees, shrubs, hardy bulbous plants and creepers are commonly used in this type of garden. This can be created on waste land or fallen land which fulfills the aim of utilization, beautification and production.

Formal Garden

Informal Garden

Plate 3

Informal Garden

Wild Garden

Plate 4

CHAPTER 3

Garden Types and Designs Contemporary View

Modern garden designs are classified on the basis of utility, need, location and materials used. Nowadays, garden types are focused on issues contributing towards ecological balance, biodiversity, environmental moderation and sustainability as well as personal preference and taste. This has given new perspectives to roof gardening, rock gardening, home gardening, kitchen gardening, green belt development, traffic islands, avenue plantation etc. Further there is incorporation of alternative energy sources like solar systems, water harvesting, etc. A garden could be for entertainment or a place to relax for some people and for others, a productive kitchen or herbal garden is more important. Some common garden types are given below, where in it is also possible to combine elements of two or three different types to create something entirely different.

The Cottage Garden

Cottage garden based on free garden style is unique in itself as it follows simplicity, tradition, blend, colour and utility. As the word denotes, cottage garden is created along the cottage depicting a blend of human and nature. English in origin, the cottage garden depends on grace and charm rather than grandeur and formal structure. The garden must be appealing, energetic, have a romantic feel, free-flowering and satisfying, filled with the plants, flowers, scents, and colors of personal choice. The cottage garden is designed to appear effortless and natural, rather than contrived or pretentious. Instead of artistic curves, or grand geometry, there is an aesthetically designed irregularity and a perfect blending of cottage and garden. Borders can go right up to the house, lawns are replaced with tufts of grass or flowers, and beds can be as wide as needed.

Instead of the discipline of large scale color schemes, there is the simplicity of harmonious color combinations between neighbouring plants. Native plant species and those adapted to the local climate are usually selected for these gardens. Dense planting with herbaceous perennials, colorful annuals, climbers, fragrant roses with a mixture of herbal and vegetable plants constitute these gardens. Some popular annuals include peony, cosmos, foxglove, snapdragon, pansy, bachelor's button, columbine, bleeding heart, and hollyhock. The earliest cottage gardens were more functional with an emphasis on vegetables and herbs, along with some fruit trees, perhaps a beehive, and even livestock with flowers filled in between the spaces in between. Modern day cottage gardens include countless regional and personal variations of the more traditional English cottage garden and ornamental grasses along with lush green effect.

Traditional or antique looking materials are often used in paths, arbors, Pots, ornaments, and furniture and fences. Wooden fences and gates, paths covered with locally made bricks or stone, and arbors using natural materials all provide a feel to a cottage garden. Important considerations in creating cottage garden is natural effect with homely feel and personal touch with utility criterion. Comfy traditional seating furniture (wicker, or painted metal shellback chairs), Weathered wooden fences, arbors, and gates, plants with romantic feel i.e having soft and delicate texture and pastel shades and fragrance, curving pathways, paving materials include wood chips, stone, old bricks, and flagstone, urns with old feel and colour etc. add the feel of cottage garden.

Family Garden

Family garden is a garden meant for family get-together and to serve as utility for the family members. The purpose of family garden includes a place to relax, a small tea break area, a children playing area, to enjoy home grown fruits and vegetables and physical exercise to the family members. Usually fairly formal in design, this type of garden provides safe space where children may enjoy eating and playing. Some features of family area include a patio or terrace, lawn (if for large areas only), paved area (for dining and seating arrangement), swings or benches, flower beds and separate area for vegetables and few fruit trees. Usually there is separate playing area for children and for vegetable growing. The front area is highly ornamental with good display of hanging baskets and container plants. Family garden may vary in size and according sections could be restricted and provided if the space permits. Further, selection of fruit trees can be done on the basis of availability of space and light. However, fruit trees like banana, papaya, pomegranate requiring less area may be planted. Vegetables like tomatoes, chilli, brinjal, etc and leafy vegetables like coriander, spinach, fenugreek, mint etc can be easily cultivated. Further, if the space is limited,

Formal Outdoor Living room can be created with such constructions as a patio, law walls, steps and perhaps a raised pool, this type of garden may have plenty of plants in containers but little other planting and probably no lawn. It requires minimum maintenance. In case of good availability of space, Open-place garden can be created possibly on ground sloping away from the house, this type of garden generally require low-maintenance and is ideal for passive recreation, with good views of the surrounding landscape. It may be terraced and have space for growing some fruits and vegetables. In addition to this, on the basis of type of building, the garden is classified as

a) *Front Garden:* This type of garden is for the area between the house and the street road, with the aim of decorating the house with a pleasing entry in house. Fencing is necessary to locate area which is generally small. For this, small trees or shrubs, potting plants near steps, arches at gate and lawn are the best features.

b) *Garden for Single House:* Generally the house is at center surrounded by garden and so the aim is to adjoin a separated house with its surrounding. The basic principle is to use native and local plants which harmonize with its surroundings. Evergreen hedges, terraces, drives and paths, pond, etc. are arranged in front side while kitchen garden should be placed in backyard. The playground for children should be located in side areas.

c) *Gardens for a Row House:* Usually plots of lands are small and narrow for the house placed in rows. It looks nice when narrow passage is divided into small square plots by hedge or border or dry walls. This creates variety and efficient land use. Tall, narrow plants should be avoided as they give appearance of more narrowness.

d) *Gardens for Flat in Apartment:* This type of buildings can be decorated with the principles of roof or balcony gardens and also by using indoor plants. Generally potted plants are used. Miniature gardens are another option or room decoration. One can use any possible features in roof or balcony gardens.

Secluded Garden

Such type of garden is suitable where privacy is a priority and exclusively meant for personal use. Such garden can be seen from above from the window or balcony. In general, garden should convey the feeling of retreats from the hectic and fierce world of outside. Hidden gardens were an interesting feature in sixteen century English gardens. A desire for intimacy, seclusion and privacy has highlighted the importance of this style. Seclusion offers the luxury of interesting and imaginative landscape which is isolated and free from any sort of disturbances. This type of design usually incorporates a pergola or arbor and plenty of climbing plants. A secret garden should be reached by a winding mysterious path under overhanging trees or discovered by approaching through an opening in wall or a tall hedge. A few backyard ideas that work well with the extra privacy seclusion offers are water features, dedicated backyard gardening and multifunctional family spaces. A tall fencing surrounding the area to be landscaped ensures better privacy and security. Tall and columnar canopy trees like *Casuarina*, *Polyalthia longifola*, can be used for screening purpose to enhance the air of secrecy and separateness in the garden.

Secluded garden could be used for different purposes like a couple enjoying romantic isolation, or doing gardening operations or for recreational activities such as swimming, tennis or basketball as well as for outdoor dining with being seen by trespassers or disturbed by outsiders. It is a kind of inward looking secret romantic garden where emphasis should be given on scent, as scents are strongly evocative. Privacy an important feature of secluded garden, brings with it magical elements of mystery and surprise, which produce a feeling of inner space besides adding interest to the composition as a whole, which is impossible to be achieved if everything is seen at a glance. Thus, the important features for secluded garden are often specific to these purposes that include seating arrangement (swing, chair or dinning set), water feature (a quiet pond or a swimming pool), fragrant and night blooming flowers like *Nyctanthes*, Night queen, etc along with nice carpet lawn or clean paved area. Sitting should be on a clean paved area. Groundcovers and a mid height tree at the back side of the sitting arrangement soften the geometric lines of the paved seating area. Creating water features and displays adds movement and depth to a backyard. Bird baths, fountains and water features may add liveness and mobility in the secluded gardens.

Often elegant and classic statue displays with evergreen surroundings add to the charm of these gardens. Romantic feel can be created with pink and pastel shaded flowers of roses, daffodils, hydrangeas and blooming climbers like *Clerodendron splendens*, *Adenocalyma, Quisqualis indica* and fragrant climbers like *Jasminum auriculatum,* Echites, rose climbers etc specially adjunct to the seating arrangement. Formal layouts and trimmed edges are not needed. However, spring bulbous plants like amaryllis,day lilies, calla lily, spiderlily and herbs like lavender, rosemary, honeysuckle, primrose and wild flowers accentuate the effect. Even garden lights play important role for the secluded garden, provided these are installed according to the garden demand like low and dim hidden lights for romantic feel, bright lights for indoor sport activity and medium lights for dinning purpose. Near the base of the fence, lights provide a safety element and ambience for nighttime gatherings. Some simple effects can add charm and detail to the secluded garden like wide cap piece over the fence and decorative latticework sculpture hanging on it. A low stone bench, with river rock collected on top and at the base, offers an additional seating spot.

Terrace Garden

A terrace is a raised space of ground constructed around a dwelling house or at the corner of a garden or on the sides of a hill. When this terrace is used for some sort of gardening this is known as Terrace Gardening. This type of garden is basically meant for a place of leisure and pleasure. A terrace garden is generally construed just in front of the house from where a view of the whole garden can be obtained. A terrace garden is very often referred to as the outdoor living of drawing room. To make it more cozy and private, some sort of walls are to be constructed on the sides or ends. A low brick or stone retaining wall may be built while on the top of it about 90-120 cm tall wall of lace patterned concrete blocks may be built which will provide privacy but will not stop breeze or obstruct light. Since a terrace is used as an outdoor living room for relaxing, the area should always be nearly dry under the foot. For this particular reason, most people favour a paved terrace. But grass is also suitable for terrace gardens, which hardly receives rough usage. To facilitate easy maintenance, for more comfort as a sitting place and to keep the place dry paving is done with local stones, flag stones, brick, concrete, wood, gravel and mosaic tiles. A gravel paving will be the cheapest but it subsides with the pressure of heavy furniture and children injure themselves while playing. A combination of brick and stone paving or a wood and stone paving, etc., can also be tried. As a general principle, the paved surface itself should have very little plant growth. Wide joints of the paving allow more growth of grass which is difficult to clip and requires more time in maintenance than narrow unplanted joints. Moreover, such joints are hazards for the ankles of the playing children as well as adults. Too many creeping plants in the joints restrict the movement and reduce the sitting space which is the prime necessity in a terrace garden. A few low growing creeping plants such as *Portulaca, Lantana sellowiana, Thymus subphylum* and *Veronica repens* can be recommended for planting at the wider joints. Some small specimen dwarf trees or shrubs can be planted amidst grass in the circular, rectangular or square beds left out. Some interesting and attractive potted plants in tubs and bowls can also be arranged artistically especially in completely paved terraces. Hanging baskets can also be displayed in the terrace. Plants can also be displayed in well designed plant stands. A lily pool, sundial or birdbath or a stone sculpture may also be constructed in suitable garden. Chairs and tables may be arranged in places for resting and comfort.

Paved Garden

A paved garden, if properly laid, can be a very attractive feature of garden. There are some specific plants, which adapt themselves well to a paved garden. These should be dwarf in nature and a considerable amount of water and tear from shoes of different weights. Ordinarily a paved garden is meant for walking, although not very frequently and hence the interstices of the paved garden should be planted sparingly.The stone selected should not be less than 25 cm in diameter. For paths, rectangular stones are more suitable, whereas for circles and squares irregular stones of any shape or size can be used. Between two stones a gap of about 4-8 cm is left and the stones are laid in an informal pattern. The boundary of the paved garden is usually supported by some rock work, supported by stones. The plants grown to cover these are allowed to grow a little inside the paved garden in an informal manner to give a more natural effect.

Planting should be done with discretion and over-planting should be avoided. The plants should be placed irregularly either in small groups or on isolated patches. A dense planting will look unnatural and will have very little appeal. Some of the plants suitable for a paved garden are *Achillea rupestris, A. tomentosa, Alyssum montanum, Coreopsis, Dianthus deltoids, Euphorbia splendens, Geranium pylzowiamum, Gypsophila repens, Hymenatherum tenuifolium, Lantana sellowiana, Portulaca (Perennial), Setcreasea pallida, S. purpuria, Verbena crinoids, Vinca rosea, Viola cornuta, Zinnia linearis*, etc.

Roof Garden

A roof and balcony garden is an urban feature. Roof gardening refers to gardening on the roof which may be in form of tubs or troughs or by filling up media on the terrace floor. The growing demand of greenery and soft structure among the concrete jungle of modern housing developments, strip malls and business complexes has given rise to roof gardening concept. Rooftop gardens offer a place to produce fresh fruits, vegetables and flowers and help to reduce the ecological footprints left by ravaging the earth to build things we can't live without. Rooftop gardening is an attractive and energy-saving alternative to a conventional rooftop. Plants have the ability to reduce the overall heat absorption of the building which then reduces energy consumption. A roof garden is an area that is generally used for recreation, entertaining, and as an additional outdoor living space for the building's residents. It may include planters, plants, dining and lounging furniture, outdoor structures such as pergolas and sheds, and automated irrigation and lighting systems. A roof garden reestablishes the relationship between humans and nature that can be lost in urban environments. It is a good means of transforming useless areas into lush, productive gardens. Rooftop gardening also allows rainwater to be retained and reduce urban runoff that otherwise would collect pollutants and empty into sewers. A rooftop garden filters and moderates the temperature of any water that is released to the sewer. The building and surrounding area's aesthetics are enhanced and property value also increases through roof gardening. The important considerations before developing roof gardens include roof top load capacity(*Extensive green roof systems weigh about 20-34 lbs/ft2 and Intensive green roof systems weigh about 80-150 lbs/ft2*), exposure to elements (wind, sun and heat), Access and safety (Enclosures with. guards, railings, parapets, walls around rooftops, terraces, and balconies), roof properties and drainage, cost for laying out of roof garden (based the structural analysis and design assistance), maintenance etc. The temperature of roof can actually be 5°C higher than the surrounding land. Shade net (50%) may be used to reduce the sun exposure. Other applicable issues might include possible upgrading of drainage and water-proofing requirements, exiting requirements: types of exits allowed and number of exits required and possible requirements for fire alarms, exit lights, emergency lighting.

Construction and design: A complete green roof consists of many layers that impersonate the conditions found in nature, creating an environment suitable for plant growth. Typically, the cross section of a green roof begins (starting from the bottom) with an insulation layer, a waterproof membrane to protect the building from leaks, and a root barrier to prevent roots from penetrating the waterproof membrane. For the waterproof membrane, look for products that can withstand the effects of acids released by some plant roots. A drainage layer, usually made of lightweight gravel, clay, or plastic is next. The drainage layer keeps the growing media aerated in addition to taking care of excess water. On top of the drainage layer, a geotextile or filter mat allows water to soak through but prevents erosion of fine soil particles. Finally, the top layers consist of growing media, plants, and a wind blanket. The growing media is lightweight material that helps with drainage while providing nutrients to the plants. A wind blanket is used to keep the growing media in place until the roots of the plants take hold.

For water supply, the gardener would generally depend upon the source of supply. Therefore, he will have to assess how much water supply is available and whether he needs to restrict the number of plants and also have to choose the time for watering the plants according to the availability of water or will have to provide a small tank or tub to store water for the garden. The investment on this account will include: Fitting up a tap on the water tank if such a tank exists; Plastic pipe of 1.25 cm (1/2 in.) diameter, 1½ times the distance between the tap and the farthest point on the terrace. The additional length is provided as the pipe is seldom taken in a straight line from the tap to the plant. It has to avoid other plants and follow some curves of the intermediary roof walls; and a tank or tub, if necessary, for storage of water.

Some guidelines to be followed for roof top gardening include use of lighter growing media as perlite, vermiculite, peat moss and coconut husk fibre, placement of the largest planters and containers at building's structural columns locations and spread the smaller, lighter ones about equally, Selection of hardy plants that can bear harsh climate, provision of shade with 50% shade net enhances the selection capacity of plants for roof gardens, provision of thick layer of mulch to insulate and shade the soil, drip irrigation system and provision of staking to the plants.

There are two types of rooftop gardens, depending on the structural design of the roof, one is the extensive rooftop garden, often inaccessible and the other is the intensive rooftop garden, which is accessible to people. An extensive green roof system is generally less costly than an intensive garden. Either kind of green roof system can increase the useful life of your roof by about 50% over a conventional roof because the green roof system layers protect the "hard" roof from exposure to harsh weather. Extensive rooftop gardens are low maintenance lightweight gardens. Depending on climate and the amount of rainfall, one can grow a variety of hardy grasses, wildflowers, mosses and sedums. Intensive rooftop gardens are little costlier and high maintenance gardens that allow for a more diverse plant selection such as perennial flowers, trees and shrubs (all of which can remain in containers over the winter) and the potential to grow food. A stronger roof structure is required due to the added weight of people accessing the garden as well as higher soil and container weights, decking and trees all adding to the weight impact to the roof. Hardy or indigenous plants, having shallow root systems should be selected for these gardens. Choose plants that have thick leaves with hairy or waxy surfaces as they tend to be stronger and lose less water to evaporation like Cotyledon, Euphorbia mili, opuntia, etc. In flowers, a combination of flowering shrubs, climbers, foliage plants, coupled with a few annuals of perennial and bulbous plants would be a choice combination. There is excellent scope for cultivation of 'Bonsai'. Hanging baskets look charming. Climbers can be rained on the shade-frame and would cover it beautifully. Vegetable plants like carrots, radishes, potatoes, strawberries, onions, lettuce, red beets, cabbage, broccoli, cauliflower, spinach, Swiss chard, mustard and collard greens and peppers can also be selected. A few herbs like mint and coriander, lettuce, onions for salad, pungent chilies and Chinese radish can be grown. In the fruits 'beauty seedless' grape, '*Karaunda*' and 'Kagzi' lime are suitable plants for roof gardening. These can be grown in containers of about 45 cm (18 in.) diameter.

Rock Garden

A rock garden is a pleasant combination of flowering and ornamental plants blooming and flourishing among rocks. In the most simple terms, rock gardening is a naturalistic style of gardening that uses rocks, large or small in some way. This type of garden that features extensive use of rocks or stones, along with plants native to rocky or alpine environments. There are rock garden plants that need to be grown in close proximity to rocks in order to thrive and those that are simply attractive when grown with rock accents. Many rock garden plants are compact in size, making it an ideal style of gardening for those who have limited space. A rock garden should be situated in an open and sunny situation. It should be away from large trees and nearby hedges as their roots may interfere with the rock plants. The site should be well drained and if natural drainage is not there, some artificial means should be provided.

Rocks for rock garden should be of local origin, porous, and have a weathered look. Limestones from the quarry can be used as they weather quickly. Tufa, a calcium rich rock used by rock gardeners to grow lime loving plants.Stones of uniform size having a diameter around 60 -70 cm should be selected and a few stones should be as large as can be handled without much difficulty. The placement and selection of rock is an important factor. The planting soil is a combination of garden soil, crushed rock and river sand to create a loose matrix that breaks apart easily and allows easy penetration of oxygen and moisture. A rock chip mulch is often used in rock gardens. This inorganic mulch allows water to drain quickly away from the crown of the plant, allowing the crown to dry quickly and reducing the risk of disease. Many rock garden plants thrive in low fertility soils.

Proper drainage is the key to successfully growing rock garden plants. Plants grown in rock gardens require excellent drainage for healthy growth. They are often grown on a slope or in a raised or mounded bed where irrigation or rainwater flows quickly away from plant crowns and roots. The water requirements of rock plants vary depending on factors such as how deeply their roots system extend and whether they are from a dry summer environment such those found in Mediterranean regions. Well established, deeply rooted plants often require very little water. Plants grown in containers will always require regular watering. Rock garden may include Containers and Troughs gardening, Raised Beds, Rock accented garden, berms style (raised garden with informal edges), or Hillsides, sloped or terraced rock garden. Sloped sites are naturally more freely draining than level sites. If additional drainage is required, the native soil can be amended. Large scale rock accents can create a sense of drama. Slopes can be terraced, if desired, to create relatively level planting areas. Water features near the rock garden provide focal points in a rock garden, enhancing its naturalistic feel along with the sound of running water add another element of interest.

A rock garden is a permanent feature of the garden and it should be laid out after careful planning. Before placing and setting the rocks the mound should be allowed to settle for some time and firmed by beating, if needed. The placement of stones will be such that it gives protection to the roots, and, hence, placing of stones too perpendicularly will not serve the purpose. The stones are placed in a slightly slanting position gently leaning backwards, not forwards so that rain water flows towards the plant roots. No rock should hang over a plant as it will deprive the plant below of rain water. The rocks should slope in the same direction with the best faces visible, so that they represent a natural stone strata. Irregular in shape pockets are created in between stones varying in size from 15 cm to 60 cm across for plants to be grown, should be and constructed in such a manner that the compost inside is not washed out. A broad path, about 45-60 cm, winding through the rock garden between the gorges and the valleys maybe provided to make all the parts of the rock garden accessible. The rockery should be planted with grown-up and well-established plants. The first thing that is to be kept in mind is that plants will fill up the space as they grow. A pendant plant hanging down a projecting stone or a prostrate plant rambling along a large stone presents a pleasant view. For dry rock garden different cacti like Opuntia, Cereus, Mamimillaria, etc and succulents like Agave, Aloe, Brayophyllum, Euphorbia splendens, Furcraea, Gasteria, Haworthia, Kalanchoe, Cotyledon, Sansevieria, Sedun, Sempervivunt, Stapelia, and Yucca can be planted in the rock garden. In a small rock garden large species of Agave, Furcraea, and Yucca should not be planted. Shrubs like *Adeneum obesum, Asclepias curassavica, Asystasia coromande liana, Azalea, Brya ebenuts* (near water garden or stream), *Calliandra inaequilatera Callistemon lanceolatus* (near water garden), *Casia alata, C. laevigata, Cytisus** sp., *Daphne** sp., *Duranta phmieri variegate, Euonymus** sp., *Jatropha podagrica, Juniperus chinensis J.horizontalis, Lantana Sellowiana, Muehlenbeckia platyclados, Ruellia* sp., (semi-shade), *Russelia Juncea, Samchazia nobillis variegate, Vaccinium* sp. Etc are also a good choice for rock garden.Some hardy ferns belonging to the genus *Drynaria, Nephrodium, Nephrolepis,* and *Poly-podium* can be grown in the shady portion of the tropical rock garden. Bulbous plants like *Amaryllis, zephyranthus*, calla lily etc also look amazing in between rocks. Flowering Annuals at different pockets in the rock garden may be added to bring colour like Antirrhinum(dwarf), *Bellis perenuis, Brachycome iberidifolia, Browallia elata,* Candytuft (dwarf), *Dianthus chinensis* (dwarf), *Hymenatherum tenuifolium, Mesembryanthermum criniflorum,* Phlox (dwarf), *Salvia splendens* (dwarf), *Verbena hybrida, Viola tricolor hortensis* (Pansy) and *Zinnia linearis.*

Moon Garden

The moon lit night in its shadows and glistening stars, holds secret tales and mysteries that are hidden in sunlit day. Thus, a moon garden is a garden to relax at night and enjoy the beautiful sky, feel the blowing breeze and smell the fragrant air in peace. There is something quite magical about a garden at night. Fragrant flowers, specific plant canopies and shape along with a comfortable bench invite visitors to sit and enjoy cool evening breezes. Light colors and white take on a new glow, and many blooms appear to float because the green stems and leaves fade into the darkness. Lighter colors of variegated plants become more pronounced in the evening. Sound is another element to consider. At sunset, when temperatures cool and breezes whip up, there's mystical experience with the swishing foliage of grasses, bamboos, and tree leaves as they flutter in the wind with soft fragrance touching and floating with the breeze. One of the most important components of a moon garden is sitting arrangement as it need to be comfortable to enjoy night beauty at a long stretch under open sky. Sitting arrangement need to be comfortable, it may be a restful chair in the middle of the garden or a bench along the perimeter with silent and soft meditative surroundings. Seats and benches next to the water fountain, with white flowers in bloom, jasmine and night queen filling its sweet fragrance is a unique way to entice special guests and feel the garden's charm. There should be different places to sit, so that your guests can enjoy your garden. Water features are also important as they give cool and soothing effect, hence slow flowing or floating water falls add grace to the moon garden. Small silent pond having white blooming lotus along with soft and elegant low light add serenity to the meditative environment of moon garden.

A very important factor is lighting in moon garden, as it is meant to be enjoyed in open at night. Lights should be of low intensity at low height, sodium lamps may installed at greater heights, LED and solar power light rocks also look elegant. Also direction of light is important as it should be from top towards down facing. Selection of plants for moon garden is very important. White flowers, light colored or silver flowers, luminescent (pale yellow) flowers glow in the moon light. Night-blooming flowers are among the showiest blooms in this garden like moonflowers, four-o'clocks, and angels' trumpets, add their own unique qualities to the garden; their fragrances attract night pollinators. Fragrant flowers like Night queen, *Murraya exotica, Nyctanthus arbortristis*, Jasmine, fragrant white or pink rose are highly preferred. Any type of the plants that have the striped foliage or have a texture to them look really good under the moonlight. any grass or shrub that has a light foliage (dusty miller, or a light colored flax), anything that has a light color is also perfect for your moon garden. Bulbous plants like amaryllis, daffodils, lilies, spider lily white tuberose and Casablanca lilies with their incredible fragrance add to the special charm of moon garden. White flowering chrysanthemums are intoxicating. Big white dahlias just reflects the moon and looks gorgeous. Night blooming cactus also looks enchanting.

Bog Garden

Bog garden refers to a garden being created in a marshy with water loving plants. It adds to the water feature specially near a pond or pool edges or in a low lying area that behaves like a water catchment area. It often like a transition or a linkage between water body and other parts of the garden. Even sideways of natural streams are perfect location for bog garden. This may be possible naturally or may be achieved by creating a large, shallow depression about 30 cm/12" deep which can be lined with punctured polythene at different poistions,. Water percolates easily through the wholes and easily provides a muddy area . The holes in the polythene should be covered with gravel to avoid clogging, followed by adding soil mostly clayey type with some farm yard manure and water. The edges of this artificially created area should be banked up with soil forming a high ridges, thus separating it from water body so that the water in pond or pool doesn't get muddy. Bog garden has its own beauty as it increases the number of species to explore in the garden, giving a unique look to the garden and adds beauty of the water body banks and surroundings. Moreover, it provides a thrill of nature's beauty existing in a wide range and under all kinds of conditions. Another benefit of bog garden is that a marshy area in your landscape area can be well utilized for creation of bog garden. Some types of grasses like *Typha angustifolia, , Arondo donax, Colocasea esculentus, Cyperus, Hosta, , Mentha acquita, Pandanus odoratissimus, Primulas, Cannas, Iris,* lilies like *Hymenocallis, Hemerocallis, Zephyranthus*, many ferns specially *Osmunda, Onoclea, Zantesdeschia aethiopica, Troillus*, Sag as well as the huge leaved *Gunnera manuculata* and *Rheum palmatum* look highly elegant and can be well grown in bog garden.

Beside a bog garden a raised deck or platform or bridge over it may be provided that facilitates adding more visual point for enjoying the bog garden with ease. Even rocky path adjoining bog garden also looks natural and elegant, besides providing accesiblility for better visibility of bog garden.

Sacred Gardens

Plants are considered as sacred since, ancient times and we Indians also have the tradition of 'Tree-worshiping'. Tree cults, in which a single or groves of trees have been worshipped, have flourished in India throughout history. The various mythological and practical reasons for considering plants as sacred are numerous *viz.*, 1) its close association with a the divinity like - *Bilva* tree (*Aegle marmelos*) with Lord Shiva, Neem tree (*Azadirachta indica*) with *Mariamman* and *Kadama, Tulsi* (*Ocimum sanctum*) with Lord Krishna, provision of shelter and nurture to object of worship 2) having origin from bodies or limbs of Gods like *Rudraksha* tree(*Elaeocarpus ganitrus*) from the tears of Lord Shiva, 3) Occurance of specific sacred act in their proximity, major role in the local ecology. Sacredness was associated with plants to realize people the importance of plants that would encourage them to grow more plants with respect and to shun people from plant destruction. It is also believed that sanctity of the plant material is hidden in its radiation of bio-plasma i.e '*Aura*'. Our saints meditated and were enlightened under particular trees and also gave their preaching under the tree shade. A Garden based on "Sacred plants" as per Indian mythology is referred as "Sacred garden". Sacred garden is more like an unconventional natural temple without an idol but plant species providing physical and mental solace. Some plants are called sacred as they are closely associated with deity like Bilva tree with Lord Shiva and Kadam tree and Tulsi plant with Lord Krishna. Some of the traditional sacred gardens include *Shiva Panchayatana,* Nandan van, Rashi van etc. Even in modern days, such sacred gardens are constructed surrounding temples.

The Shiva Panchayatana, is one of the favourite forms of Shiva worship in which the lord is seated surrounded by Vishnu in the northeast, Sun-god in the southeast, Ganesha in the southeast and goddess Ambika in the northeast directions. This concept of the Shiva panchayatana, can be created by growing the plant species associated with the members of the panchayatana, like bilva (*Aegle marmelos*) and drona (*Leucas cepheloites*) with Lord Shiva, ashwatha (*Ficus religiosa*), tulsi (*Ocimum sanctum*) with Vishnu, khadira (*Acacia catechu*) and Cyanodon dactylon with Lord *ganesha,* karaveera (*Nerium odorum*) and Svetarka (*Calotropis procera*) with lord Soorya. *Nandana Vana* dedicated to Lord Krishna is a kind of celestial garden as portrayed by our ancient writers. Kalidasa, the famous writer, has mentioned ten trees including aswatha (*Ficus religiosa*) in "Mangalashtaka" where he signifies this evergreen and ever flowering garden full of fruit yielding trees as a harbinger for the well being of the world at large. Poet Ashvaghosha described about this garden in his works. Accordingly, this type of garden consists of the tree species *viz.*, aswatha (*Ficus religiosa*), Vata (*Ficus benghalensis*), chandana (*Santalum album*), mandara (*Bauhinia purpurea*), kalpadhruma, jamboo (*Syzygium cumini*), nimba (*Azadirachta indica*), kadamba (*Anthocephalus cadamba*), choota (*Mangifera indica*), sarala (*Pinus roxburghii*). The pooja of *Navagrahas* (eight planets of the Hindu almanac with the Sun situated at the centre) is one of the most popular forms of worship found to prevail in many parts of the country. *Navagraha* garden is the garden which is designed by planting of particular plant species signifying particular planet in the same position of the solar system like *Ravi* (sun)-svetarka (*Calotropis procera*),*Soma* (Mars)-palasha (*Butea monosperma*), *Mangala* (kuja) (Mars)-khadira (*Acacia catechu*), *Budha* (Mercury)- uttarani (*Achyranthus aspera*), *Guru* (Jupiter)-ashwatha (*Ficus religiosa*), *Shukra* (Venus)-atti (*Ficus glomerata*),*Shani* (Satturn)-shami (*Acacia ferruginea*), *Rahu*-doorva (*Cynodon dactylon*),*Ketu*-darbh (*Saccharum spontaneum*). *Rassi vana* refers to a garden comprising of plant species signifying rassi 'zodiac signs'. Thus, it is based on the twelve Hindu Zodiac signs and planets ruling them. Particular plant species are identified to represent as the lord (based on planets) of specific zodia signs. Such species are selected and raised in the garden to reflect the effect of 'rassi vana' like Mesha (Aries)-raktachandana (*Ptercocarpus santalinus*), Vrushabha (Taurus)-Saptaparni (*Alstonia scholaris*), Mithuna (Gemini)-panasa (*Artocarpus heterophyllus*), Karka (Cancer)-palasha (*Butea monosperma*), Simha (Leo)- padari (*Stereospermum chelonoides*), Kanya (Vigro)-amra (*Mangifera indica*), Tula (Libra)-bakula (*Mimusops elengi*), Virschika (Scorpio)- Khadira (*Acacia catechu*), Dhanu (sagittarius)- ashwatha (*Ficus religiosa,* Makara (Capricorn)-Shimshapa (*Dalbergia latifolia*),Kumbha (Aquarius)-Shami (*Acacia ferruginea*), and Meen (Pisces)-Vata (*Ficus benghalensis*).

Home Garden

A garden in a home not only adds to its beauty, but also enhances the real estate value. The home along with its surroundings depicts the outward expression of the personality and individuality of the owner of the home. Basic Principles of a home garden include neutral background of a home ground whether a wall, tall trees or a hedge, with some contrast effect with variation in form, texture or colour, proper balance and proportion between different components of a home garden, with open centre either with lawn or pavement or sand along with repetition or duplicating some features to achieve rhythm, balance and unity in a garden. It is important to first ascertain the purpose for which the garden is needed that varies according to individual's taste and preference like an outdoor living room (i.e., drawing room) with long stretch of lawn and terrace (raised area), a fenced-in- playground, a showpiece with a collection of rare plants or a place for producing cut flowers, vegetables and some fruits.

If the area available for gardening is large enough, it should be divided into different sectors like 1) Approach or Public Area- It is the front area that should not be overcrowded with large trees, may have a spacious lawn with a specimen tree (if the area is large). A herbaceous perennial border, a few annual beds and foundation planting (near the doorway) can be included in this area.

2) Work or Service Area- it should have privacy and preferably be situated at the back. It includes the kitchen garden, compost bin, nursery, tool-shed and garage. Children's play area with swings etc. can be situated here. The area is to be screened off by a thick hedge or a row of bushy shrubs. The service area should be screened with light creepers (e.g. railway creeper) on a trellis, or an ornamental hedge (e.g. *Eranthemum*). If space permits, a tennis court or badminton court may be included. 3) Private Garden Area or Living Area-It is the outdoor living area of the house, where people sit out in winter to enjoy sun or rest in summer under shade or during evenings. This area should be easily approachable, visible form living room and should be screened from unsightly objects for privacy. This area should include shaded sitting spot such as a tree, or an arbour with garden benches or swing. There should be stretch of lawn with shrub border or a few annual beds, fragrant flowering shrubs, creepers or a rose garden if desired can be included. Depending upon the choice of the owner, a number of features can be included like a shrubbery, borders, annual beds, rose garden, rock garden, lily pool (if desired), etc. An owner, who can afford it, may have a greenhouse to grow decorative foliage plants. Garden adornments like a statue, sun-dial, bird-bath, benches, fountain, etc, may be included.

The doorway needs to be treated especially as this receives more attention from a visitor. This should be designed with much care and thought. Perennial and annual flowers should be planted. If planting in the ground is not possible, pot plants can be arranged. A rose garden can be designed provided the place receives the morning sun. An ornamental light, post or Japanese lantern can be placed in a suitable place.

A small area cannot be divided into various areas (like public area, living area, etc.), however, such compounds can have (a)A small lawn or paved area, (b) One or two specimen shrubs (e.g. *Brya ebenus, Cassia biflora, Ixora singaporensis*, *Mussaenda philippica* etc.),c) places having shade should have herbaceous perennials like *Impatiens sultanii*, geranium and ornamental foliage plants.

Kitchen Garden

Kitchen gardening is the growing of vegetable crops in the residential houses to meet the requirements of the family all the year round. Vegtables may not be very colourful but they reflect a feeling of quiet charm where grown. This type of gardening has been in practice even in earlier times as reflected in from historical background be in Indian, Roman, Scotish or English. However, up-gradation of the kitchen garden to ornamental status where fruits, herbs and vegetables are grown for visual effect is a modern concept. Kitchen gardening aims at an efficient and effective use of land for growing essential vegetables for daily use of a family, it plays an important part in vegetable production. Besides being a healthy hobby benefits of kitchen garden are many of which few include availability of clean and hygienic vegetables specially Leafy vegetables like coriander, spinach, colocasia at your door step, children may be involved with the concept of 'work with play' and the feeling of sensitivity towards plants can be developed in childhood. Further, ecological benefits can be gained by the use of organic manures, companion planting, deep beds and flowers to encourage beneficial insects.

Location is the most fundamental criterion for success of a kitchen garden. As most of the work is done by the family members in spare times, the location should be in the backyard of the house close to the water tap or other source of irrigation. Drain water from the kitchen can be profitably utilized. It should never be located in the shady area of home which is generally not suitable for most of the vegetables. There should be enough of sunlight for major part of the day. The soil should be porous having good drainage and it should nutritionally rich and should have sufficient amount of organic matter with soil pH should be near neutral between 6.5 to 7. For root vegetables, the soil should be light and sandy for their proper growth. The other vegetables can grow well in organically rich heavy soils also. Thus, the soil between sand to heavy but well drained and fertile with neutral pH is preferred.

The design of a kitchen garden depends on the character of the particular piece of land, its extent, situation, etc. The design could be formal or informal as per the preference and shape of the landform. The following principles should be followed in designing the layout of the garden:

1) The layout should be such as to make the garden look attractive and allow access to all the parts. As various kinds of vegetables will be grown in different parts of the year, the land will have to be laid out in small plots with narrow and path borders. The design could be either in circular form, or square or rectangular as outer border with inner beds.

2) In homes where no space is available one can grow vegetables in pots or boxes but preference should be given to such vegetables which produce more number of fruits from individual plant, i.e., cucurbits, tomato, brinjal, chillies, etc.

3) One or two compost pits can be dug in the corner of the garden.

4) The quick growing fruit trees like papaya, banana, kagzi lime, etc. should be located on one side preferably on northern side of the garden so that they may not shade other crops.

5) Climbing type vegetables like cucurbits, peas, sem, etc, can be trained on the fences.
6) Several sowings or a succession of sowing of one particular crop at short intervals should be done to ensure a steady supply of vegetables.
7) The ridges which separate the beds should be utilized for growing root crops like radish, turnip, beet, carrot, etc.
8) Early maturing crops should be planted together in continuous row so that the areas may be available at once for putting late crops.
9) The interspaces of some crops which are slow growing and take long duration to mature, like cabbage, cauliflower, brinjal should be used for growing some quick growing crops like radish, turnip, palak, lettuce, etc.

The crops to be taken in the kitchen garden depend mainly upon two factors, i.e., size of the garden and the choice of the family. Only those vegetables should be taken which are suited to the region and produce satisfactory yield. In case the land available is large for the kitchen garden, a large number of vegetables that the family likes, can be grown. If space is limited, only those vegetables can be grown which give better yield per unit area. The cultivars should be selected according to the suitability of the region and according to the period of sowing. Tomatoes, beans, cabbage, lettuce, spinach and some root crops are desirable for small gardens. In fact, in kitchen garden one should grow those vegetables, in which freshness is of great importance from the stand point of edibility and food value. Even some edible flowers like marigold, nasturtium , damask rose as well as mustard, sweet pea, calendula, onion, etc and aromatic herbs like fennel, coriander enhance colour and aesthetic look of kitchen garden besides, adding scent to the garden. Further, proper planning of crops, well defined paths, cleaniness and neatness provide glistering effect with health benefit and becomes a visual artistic piece of land. Staking is also important in vegetable garden, making wigwams of of bamboo canes, tied round with garden twine, make splendid support for sweet peas and gourds which climb forming a nice cone. In veg garden, care is needed to neatly harvest the vegetables and later to remove the used up plants followed by frequent sowing of next season vegetables at the right time. One or two rows of staked tomato may be planted by the sides of the path. Ridges may be utilized by growing root crops like carrot, beet, radish and colocasia. Climbing vegetable crops like bean (sem) and cucurbits (pumpkin, bottlegourd, prwal, etc.) may be planted near the fences of the garden. Chilli, ginger, turmeric, coriander, etc., can be grown as intercrop in the perennial block where some quick growing fruit crops like guava, citrus, banana, papaya, pineapple, etc., may be planted.

Herbal Garden

Grow herbal plants in your garden and enjoy fresh and healthy life with natural benefits of herbal products. Herbal gardening is growing of herbal plants i.e. plants having medicinal properties in the residential areas and utilizing the fresh herbal products daily as and when required. These include medicinal, aromatic plants as well as the spices and condiments. Herbs look invariably pleasant, many of them have beautiful flowers and accentuate scent effect in the garden. Benefits of herbal garden are: 1) fresh herbs are available for use as per the requirement, 2) Clean and hygienic insect-pest free herbal leaves like curry-leaf, cinnamon, garlic, ginger, *Ocimum* (tulsi), coriander, etc. can be obtained, 3) Children can be involved in the herbal gardening and can develop the herbal knowledge since childhood 4) Herbal cure for minor ailments like sore throat, cold, cough and cuts or wounds is available at hand.

The herb garden can be designed formally in some kind of parterre or herb wheel, where dividing path can be graveled. Plants like sage, rosemary, marigold, nastrutiums, calendula, mustard, onion and thyme should be grown for colour. Flowering plants should be kept as borders. There are many herbal plant species which can be easily grown in our residential areas. Here, the common herbal plants which can be easily cultivated and have everyday use like Basil (Tulsi)- *Ocimum basilicum, Ocimum sanctum*, coriander (*Coriandrum sativum*), lemon grass (*Cympopogan citratus*), mint *Mentha viridis*, etc. Also *Aloe vera, Aloe arborescens* can be easily planted.

Besides, marigold and *Calendula officinalis* are both a beautiful flowering annual herb for antiseptic properties and as flavouring rice and pulao. Herbs like Chives- *Allium schoenoprasum*, Anise(*Pimpinella anisum*),Fennel (Variyali, sauf)- *Foeniculum vulgare*, Dill (*Anaethum graveolens*) have beautiful umbels besides herbal properties. Even bulbous plants like Garlic- *Allium sativum*, onions, ginger, can be cultivated for use in kitchen besides their medicinal properties alon with their beautiful flowers. Shrubs like Cinnamon, damas rose, and trees like all spice (*Pimenta officinalis*), curry leaf (*Murraya koeingi*) are highly aromatic and full of medicinal and culinary use in everyday life.

Family Garden

Secluded Garden

Terrace Garden

Plate 5

Paved Garden

Rock Garden

Plate 6

Moon Garden

Bog Garden

Sacred Garden

Plate 7

Home Garden

Kitchen Garden

Plate 8

CHAPTER 4

Garden Designing Principles

Garden should serve the feelings of comfort and pleasure along with soothing effect on mind and soul of the owner and the visitors. This should be the basic aim of the landscape designer along the budget. Landscape artist should never imitate another garden which has secured popularity but should develop one's own design. Every plot or site may not be even and ideal for designing a garden. The principles of landscape design work together to create the overall look of the garden. Natural undulations of a barren landform should be utilized with its advantages and also to break the monotony of leveled plain. The principles of garden designing interact with the elements of color, texture, line, form and scale to develop a unified landscape design that blends the beauty of nature with the splendor of modern architecture.

Principles of Garden Designing

1. Style

Selection and finalizing the style of the garden is the first step after seeing site, for planning a garden design. Style should be selected on the basis of purpose, types, landform, interest of owner, maintenance capacity, region (urban or rural area), building structure, etc. One has to be creative to evolve one's own style of gardening according to his budget, taste and the nature of the site. There are three basic garden styles :i) *Formal garden*: It may have a symmetrical or a geometrical pattern. In this pattern, everything is planted in straight lines. The flower beds, borders and shrubberies are arranged in symmetrical beds. Trimmed formal hedges, Cyprus, Royal palms, Ashoka trees and topiary are typical features used for the same. ii) *Informal garden*: Informal garden depicts and imitates the nature. Here, curves are more predominant over geometrical rules.

Whole design looks informal, as the plants and all features are creatively and harmoniously arranged in a natural way without following any sort of principles. However, they should not be haphazardly arranged without harmony and good look. iii) *Wild garden/Intermediate:* This style is like tree style not bounded by any rules. The main features of this style are natural shrubberies, or bushes (without pruning), grass unmoved and herbaceous perennial plants should be grown scattered in grasses as like in wild scenery.

The important feature of some typical garden styles include a) English gardens: Lawn, herbaceous border and rockery, b) Italian gardens: Heavy masonry features like flight of stairs, fountains and statues, c) French gardens: Large space and avenue plantation. d) Persian gardens: Crafted material, water canal and terrace, e) Mughal gardens: Walls, gates, terraces, running water, *baradari* and tomb or mosque, f) Japanese gardens: Ponds, streams, waterfall, fountains, islands, bridges, water basins, stone lanterns, stones, pagodas and fences.

2. Balance

Balance in the garden design creates visually pleasing effects to the landscape be it asymmetrical or symmetrical. Symmetrical balance duplicates the garden design on one side of a clearly defined central axis and repeats the exact same design on the opposite side. Each side of the design is a mirror image of the other with no variation in color, texture, or other elements. Asymmetrical balance is less rigid with natural curves and more variety in the design. The center point may not be obvious and balance is achieved through mass and weight rather than color, texture and plant types. Radial balance works in a circular pattern from a center point to produce a balanced appearance. Flowers like Vinca, Sunflowers, Tabernaemontana and elements like wheels, a well pulley or a floral or ornamental clock adds further to produce radial balance.

3. Focal Point

In every garden, there should be a centre of attraction to focus as a point of interest. It is the most important aspect of the principles of landscape design as it sets the tone and arrangement for the rest of your outdoor area. A decorative fountain, magnificent sculpture, or even a simple garden with a relaxing sitting area could become the focal point of your yard design. All elements of the landscape should work to emphasize this area. A decorative Fountain, statue, piece of sculpture, water falls or a specimen tree like Christmas tree, Bottle brush, *Cassia fistula, Anthocephalus cadmba, Cassia renigera, Delonix regia or* pleasant relaxing seat etc. can be used as a focal point.

4. Space

Space in the garden is a heart of the design and is used as a site for recreation. The aim of every garden design should be such that the garden should appear larger than its actual size. The technique of creating an illusion of more space is referred to as 'Forced Perspective'. This can be achieved by keeping open spaces, preferably under lawn and restriction of planting in the periphery. If planting at the centre is needed, the choice should be for a tree having branches at a higher level on the trunk like *Spathodia campanulata.* Another way of creating an illusion of distance is making lines converging slightly at a distance. For example, paths in a garden are gradually narrowed or the sizes of the farthest trees diminish in size. The illusion of more space in a large public garden can be created by alternately large lawns, followed by a group of trees.

5. Axis

This is an imaginary line in any garden around which the garden is created striking a balance. In a formal garden, the central line is axis. At the end of axis, there should be a focal point, although another architectural feature such as bird-bath can be erected at about the midpoint. Where as in informal landscape design axis will be zigzag or curvy and it may be a path across the design.

6. Unity

Unity in the garden creates rhythm, harmony, and balance. It refers to the central unified theme that blends with the entire property presenting a unified absolute effect. Plan every part of your landscape design carefully to ensure consistency. Choose materials, plants, and accessories that complement the theme you have selected and avoid any items that do not add to the harmony of the design. Unity in garden can be achieved from various angles: 1) The unity of style, feeling and function between the house and the garden. 2) Harmony of different components and features within the garden. 3) Harmony between the landscape outside and the garden. To achieve a unity between the house and the garden, one can train creepers on the front porch to cover the rudeness of the masonry. Foundation plantings also serve this purpose. A foundation planting means planting of bushy plants near the foundation of the house.

7. Repetition

The repeated use of one general form of plant material in group at one place is done to have a mass effect. However, it should be done carefully as too much repetition can destroy the atmosphere of an outdoor area by creating a dull appearance, but appropriate repetition of similar plants, colors, and textures bring a uniform, blended look to the landscape. Repetition can complete a garden design, tying all the parts and pieces of the plan together into a unified design that blends with the attached building structure. Examples of repetition in the garden are hedges, flower beds, borders, shrubberies, avenue planting, etc. Repetition is very important in any design to create a rhythm and balance as it introduces a sense of order to the viewer. The sizes of the masses can be varied to break monotony. Shrubs like Acalypha, Duranta, Croton, Rose, Ixora, Hibiscus, Ceasalpinia and *Murraya exotica* (*Kamini*) can be used for eliminating the monotony.

8. Rhythm

Rhythm gives a landscape design a soft feeling of natural movement through the use of natural elements and careful repetition. Groups of plants, as well as individual materials, can create rhythm within the environment by patterns of color, form, and other elements. Smooth topiaries in series, shrub or tree canopies, a formal hedge or edge, flower beds in orderly fashion all depict rhythm in a garden design.

9. Variety

It is a critical element in landscape design. Variety can be depicted in different colours, morphology and height flower beds, plant material used in edges and hedges, training of shrubs and climbers, display of pot plants, using bulbous plants. Variety can also be displayed in paving, curbs stones, paths, water bodies etc. Too little variety may lead to monotony while too much brings confusion. A unique balance between extreme creates a pleasant sense of unity with different varieties in a landscape composition. A variety of lines, forms, textures and colours are needed to create an interesting landscape. There should also be a variety of features in the ground.

Characteristics of Plant Material

A specific pattern is formed in the natural arrangement of plant parts as result of plants characteristics. It is important in garden setting to study and apply it. For example, weeping plants add delicacy, oaks and bottle palms are stately, bulbous plants look exotic while flowering trees with dazzling colours are sparkling and gay. Bright annuals add colour and change. Here, the shape and form, texture of leaves and trunk, the foliage colour, flowering time and colour, flower fragrance, photoperiodism all are the important considerations.

10. Divisional Lines

In a landscape garden, there is a necessity of dividing or rather screening a compost pit or a *mali's* quarter or a vegetable garden or special part like rose beds from the rest of the garden. The divisional lines should be artistic with gentle curves in informal (natural) garden style while it should be straight in formal design and these should also be useful. These lines should harmonize with one another. Straight lines indicate direct movement without hesitation. Inter-connecting straight lines create points at the interactions for stopping, sitting, changes of views and reflections back to the point of beginning. Rows of dwarf trees or rows of shrubs as hedge and edging plants are used to create lines. In fact areas under lawn, gravel stone or cement path, and shrubbery border have their natural divisional lines from their immediate next feature.

11. Scale and Proportion

Proportion may be defined as a definite relationship between masses which create rhythm, balance and harmony in garden. A well planned outdoor space is designed with function in mind. Closely related to the element of scale, proportion means carefully selecting materials that are appropriately sized for the landscape and its purpose. A small fish pond, for instance, would appear out of place and insignificant in a large, estate style landscape with grandiose structures but it would be ideal for a private landscape design in an average backyard.For example: a rectangle having a ratio of 5: 8 (width: length) is considered to be of pleasing proportion. There is no set rule for scale and proportion in a garden but, a simple rule is that a design should look pleasant and attractive. Scale and proportion is strictly followed in geometric garden but it is an art in the natural garden. Few more examples to illustrate the importance of scales are; narrow steps leading from a wide terrace is completely out of scale. The steps should be spaced wider, to make climbing easier and pleasant. The common practice of laying out a small rockery at the base of a large tree with small thorny specimen looks out of scale and proportion. Small tree in the middle of a large lawn or a large pond in middle of a small lawn area are also out of scale and proportion.

12. Colour and Tone

Colours and tone to be used in the garden are important as these produce deep effects on the overall garden view and feel. Three basic colours are red, yellow and blue, while the secondary colours are orange, green and violet. Red, orange and yellow are considered as hard or warm colours while blue, violet, green are considered as soft or cool. Black and grey are accepted as neutral colours. The natural and most pleasing colour scheme is the one that is found in nature like a rainbow. Colors in the garden can be used for varied purposes be it attracting attention to prominent areas, affecting the perception of distance as well as for affecting mood and atmosphere throughout the outdoor space. Colors that blend into the landscape, deep hues like black, green, and cool shades of blue, can make a home appear further away, while bright, warm colors make objects appear to be closer. Vibrant reds, oranges, and yellows convey excitement and are most appropriate in active areas of your landscape design. Further, cool shades, like blues and greens, are tranquil colors that work well in areas designed for relaxation.

The basic arrangements of colours using annual flower beds and trees in avenue plantation or shrubs in hedge and borders are

i) *Monochromatic:* It is an arrangement of different tones of the same colour either of the same species or in combination with different species. For example: Blue petunias with blue lobelia or pink antirrhinum with salmon pink candytuft. In flower border calendula used with marigold and sunflower. Rows of red or yellow flowering trees in roadside plantation.

ii) *Analogous:* It can be achieved by using closely related or harmonious colours. This may be done by using either soft or warm analogous colours. A soft or cool harmony can be achieved by using blue colour flowering plants like sweet peas, china aster, etc. whereas a warm harmony is possible by using yellow antirrhinum with orange dimorphotheca or yellow marigold with orange antirrhinum, etc.

iii) *Complementary or Contrasting:* Two opposite or contrasting colours which create striking view are called as complementary, e.g. blue and orange, red and green, etc. This can be sub-divided further into split-complementary with one dominant and two complementary colours (violet with yellow-green and yellow-orange) and triads such as red, yellow and blue. Here also the contrast may be soft or warm.

The proper colour combination with plant height and time of flowering is very important in colour and tone during landscaping. According to height the appearance of multistory can be achieved with different colours but they should harmonies with other garden components.

In a landscape garden, the permanent backdrop is the green tones of the various trees and shrubs. It is possible to layout a garden with subtle tone of entirely white or yellow flowers, but at the same time making it charming also. It is better to have masses of a single colour against a mixture of colours. For example: A bed of roses containing only a single colour of say red, yellow, or pink has a much soften tone and beauty than a bed containing a mixture of colours.

13. Texture

The surface character of a garden unit is referred to as texture. Various textures are emphasized through the use of plants and other landscape materials. In the case of plants, texture is expressed in gradations from fine to medium to coarse. The texture of the ground, plant material that is leaves of a tree or shrub determine the overall effect of the garden. The key to effective *use of texture* is creating a balance between various plant qualities in the landscape. A large amount of smooth, fine materials should be used to balance coarse textured plants and trees. However, one should gradually move through similar textures in the design for a smooth transition into each new texture.The texture of rugged looking ground can be improved by laying meticulously chosen small pebbles from the riverbeds. Examples of fine textured trees are *Gulmohar* and *Jacaranda* when in a full leaves and course textured trees are *Spathodea companulata* and *Kigelea pinnata.* The placement of all these various textures with harmony and contrast has to be achieved to get the ultimate desirable effect. Besides, some plants show dramatic effect as they change the texture and feel of the ground with their flowers or leaf fall, eg. Peltophorum, Bauhinia purpurea, Murraya exotica, Pear, plums etc.

14. Mobility/Transition

Mobility in garden symbolizes change, growth, excitement and romance. The changing of colour very sharply and contrasting from one season to another in a garden depicts movement. Transition, or sequence, gradually changes patterns in the elements to assist in easy visual movement across the landscape. Transition is a logical sequence that introduces a change in style slowly rather than all at once. For example, use medium sized shrubs to transition tall trees to sprawling shrubs. For example: to create mobility feel in the garden, trees such as Indian Almond (*Terminalia catappa*) which changes its leaf colour into striking red twice annually before falling or *Lagerstroemia flos-reginae* which also changes colour of the leaves to coppery shade in the autumn before shedding or *Madhuca indica, Kusum* (*Schleichera oleosa*) and *Ficus religiosa* having coppery red new foliage in the spring, should be planted in some parts of the garden. This improves the landscape and provides liveliness.

The movement and clustering of birds also bring life and mobility to the garden. Larger trees and birds- baths attract birds. For the smaller birds, the safety of shrubberies is needed to protect them from large predator birds. Flowering trees such as *Bombax ceiba* (Silk cotton tree) or Erythrina also attract birds when in bloom. The seasonal flowers also bring in the motion and movement of colourful butterflies. Fountains, lawn sprinkler and streams in a garden also serve the objective of movement. The lily pools should be filled with coloured fish, the movement of which will be an added attraction.

Availability of Light at Different Timings and Seasonal Variation

There are three different categories of time in a garden: i) *The daily time*; this provides different quantities and qualities of light during the course of the day. As the morning is vital for all flowers, the designer has to take this into account while planning. Noon sun creates sunburn effect on Dieffenbachia and Coleus. Most of flowering annuals like balsam, sunflower, daisy, portulaca show their real charm grow well in full sunlight. Hence, for partial shade conditions, plants need to be selected accordingly like Acalypha, Croton etc or if flowering is preferred then gerbera, Vinca rosea, Pentas, geranium as herbaceous perennials and Saintpaulia, etc as annual may be selected ii) *The seasonal changes around the year;* a good planner must roughly take into consideration the seasonal movements of the sunlight and where the shade and light are likely to fall during the different parts of the seasons iii) *Light distribution;* this is very important for proper growth of all plants. Shrubs become sickly and lanky near the trunk of the large tree, because of lack of light and possibly nutrients also.

Planting of a tree in front of window which can protect from strong afternoon sun but unplanned tree can ruin the view of the garden from inside the window. The pattern of shade caste by a fine-leaved tree on the ground or lawn can look very artistic. When Royal palm or Eucalyptus or Polyalthia are planted in a row along a path, it provides oblique bars of shade in the mornings and afternoons which cheer up walk together with adding directional effect and secure feeling. Flowering trees like *Delonix regia* (gulmohar), Jacaranda, *Cassia renigera* etc. come to full blooming with real charm only when they receive full sunlight from all sides.

Balance

Focal Point

Space

Plate 9

Axis

Unity

Plate 10

Repetition

Rhythm

Formal Landscaping

Informal Landscaping Style

Plate 11

Variety

Plate 12

Characteristics of Plant Material

Plate 13

Divisional Lines

Plate 14

Mobility

Plate 15

Time and Light

Plate 16

CHAPTER 5

Basic Elements of Landscape Gardening Landform, Plant Material and Water

Landform, water and plants form the three basic elements of garden designing. Each of these basic element is described and discussed in detail.

I Landform

"Landform" refers to three-dimensional relief of earth's surface. Thus, in simple words it is the lay of land, synonymous with, "topography". Land is defined as the surface of the earth and all its natural resources. Landform is thus 3-D view of earth's surface including all the visible feature of an area of land and landscape refers to the picture representing a view of natural inland scenery. Landform is the common module in exterior environments, a base, an axis, a thread, which ties all the physical elements, space together along the horizon or the water edge. Landforms can be classified as macro-landforms, micro-landforms and mini-landforms. Macro-landforms refers to valleys, mountains, rolling hills, prairies, desert, seashore and plains. Micro-landforms refers to mounds, slopes, level areas or elevation changes using steps and ramps. Mini-landforms refers to subtle, undulations or ripples of a sand dunes or the textural variation of stones and rocks.

Different techniques are used for explaining and representing the type of landform. These include (i) Contour lines, (ii) Models, (iii) Ratio method, (iv) Value and colour, (v) Spot elevation, (vi) Computer graphics, (vi) Hatches (disconnected lines are used) and (viii) Percentage method. These techniques are used for proper study of the landform, to identify loopholes and need to modification in order to establish proper drainage or to modify the site in order to accommodate of various elements or to create an aesthetically pleasing ground plan.

Significance and Influence: Landform has direct association and linkage with the all other physical elements and aspects of the gardens as well as the outdoor environment and activities. Landforms influence the aesthetic characters of an area, definition and perception of space, views, drainage, microclimate, land use and organization of functions at a particular site. Influence of landforms on various characters is given below:

- *Aesthetic Influence*: Landform has direct impact on aesthetic character and rhythm of the landscape. Relatively level sites like ocean tend to appear quite open, wide and expansive. It creates strong internal sense of visual continuity and unity. Hilly and mountainous areas with level landform provide a sensation of isolation from one valley to another. Size and spacing of valleys and ridges within hilly region have direct effect on rhythm. Closely spaced ridges create quick rhythm while widely spaced create slow rhythm. Level landform gives a more plain and monotonous feeling with less excitement while on hilly and sloped areas dramatic and mystic effects can be created with a little planning and artistic ideas. Even particular traditional garden style evolved on particular landform like the formal French renaissance garden were developed on level landform while the Italian renaissance developed on slope and the Chinese, Japanese as well as the English gardens preferred the hilly and sloped sites for the informal garden development.

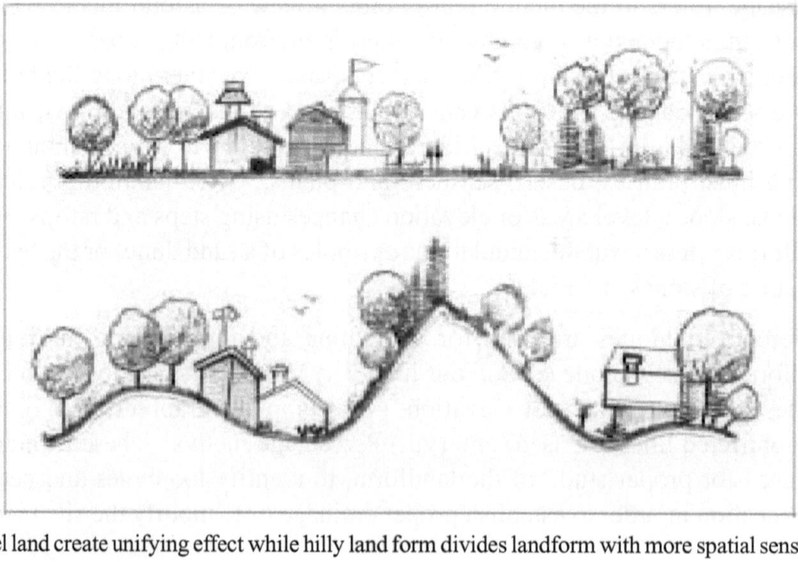

Level land create unifying effect while hilly land form divides landform with more spatial sensation

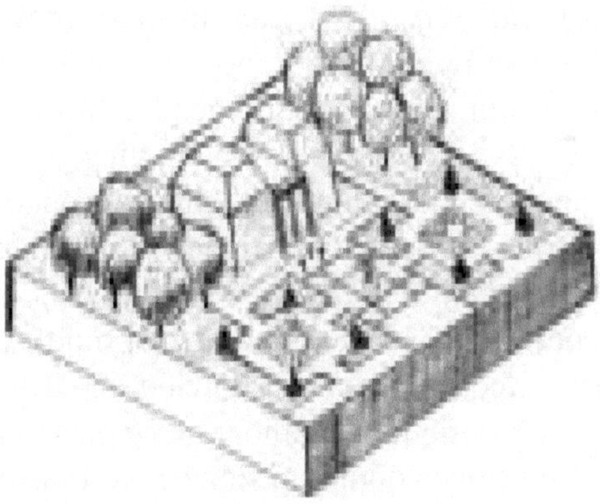

Level landform- French Renaissance garden (Formal style)

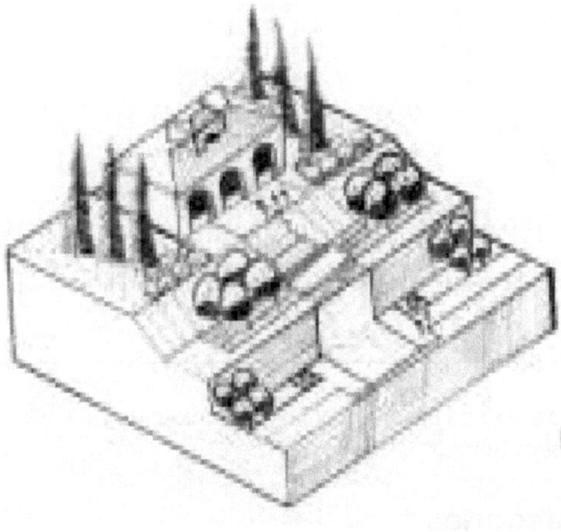

Landform with sloped site- Italian Renaissance garden (Formal)

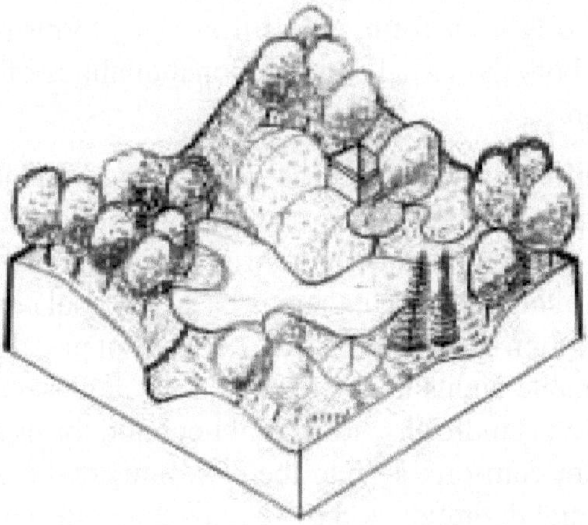

Hilly and sloped site: English garden -18th century

- *Influence on Spatial sensation*- Landform strongly influence spatial sensation at a given point. The smooth flowing landforms produce a sensation of relaxation while bold, rugged landform produce excitement and aggressiveness. *View*-Perception of view is also influenced by landform as views are closely linked to the concept of spatial definition. Different landforms create sequential viewing or progressive realization of an object. *Functional Influence*-The slope of landform influence drainage. Steeper grounds have rapid rate of runoff and swift drainage. Landforms also affects sun exposure, wind exposure and precipitation accumulation. Landform has a number of functional uses but every land use or site should have an optimum slope condition on which it operates best.

Landform influence views and pace- Level landform extends view while sloped landform creates spatial edges but blocks directional views

Landform: Categories

Landform can be categorized in numerous ways including scale, character, steepness, geological origin and forms. Classification by forms is most effective as it concerned with both the visual and functional qualities of land. Different types of landforms are:

1. *Level landforms:* Level landform is defined as any land area visually parallel to the plane of the horizon. It has slight or gentle slope. Natural, stable, restful, peaceful and at equilibrium with gravity are the special qualities of level landform. But there is lack of spatial sensation because there is no definition of enclosed space, no sense of privacy, no protection from objectionable sights and sounds and no defense against sun and wind. On the level landforms, horizontal lines and forms are harmonious elements that fit comfortably into the environmental settings. One can also create spatial definition and privacy by using different vertical and large elements. In contrast to horizontal, any vertical element that is

introduced to a level landform at center has a potential of becoming a dominant element and focal point. The flat topography is suitable for setting quite, still and reflecting sheets of water which give feeling of rest and peace. Bold forms and colour can be placed on a flat site to take advantage of eye-catching element which breaks monotony. It also provides multidirectional use so one can choose any direction to form particular point e.g. it offers scope for developing big residential society with good landscape gardening.

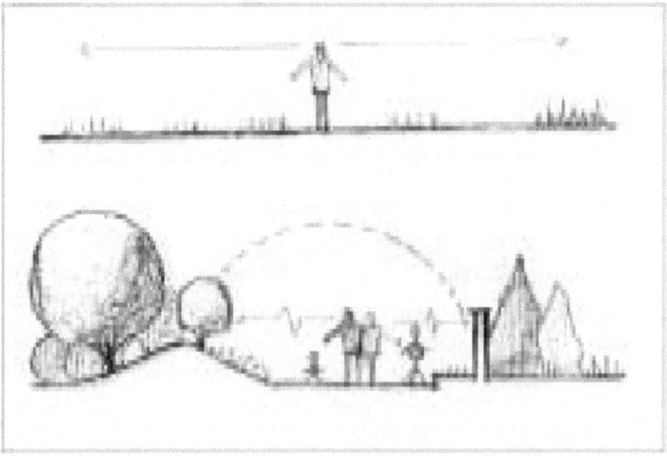

Level landform provides open and spaceous feeling, where spatial defination and privacy need to be created by sculpting ground surface and by use of other elements like trees or buildings

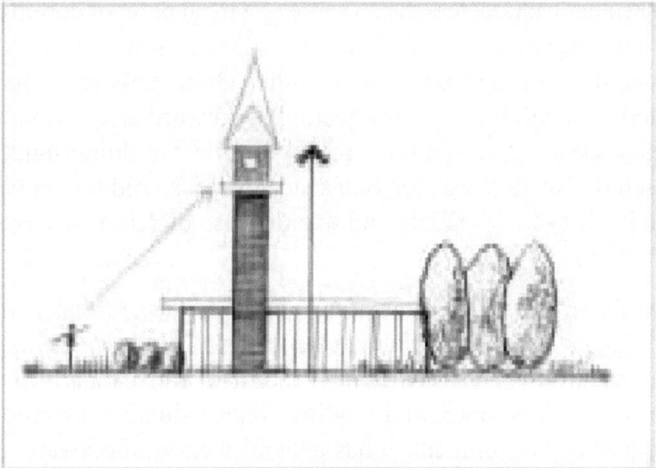

Verticle object (if single) provide a strong contrast with level landform and need to be selected carefully or supported with other verticle and horizontal elements

2. *Convex landform:* It is described as a high point of ground defined by a generally concentric arrangement of contours, for eg. mountains, hills, knobs, buttes, etc. A convex landform is a positive solid and negative space. It is dynamic, aggressive, exciting landform implying power and strength. Temples, government buildings and military camps are often placed on top of convex landform to take advantage of feeling of 'elevation'.

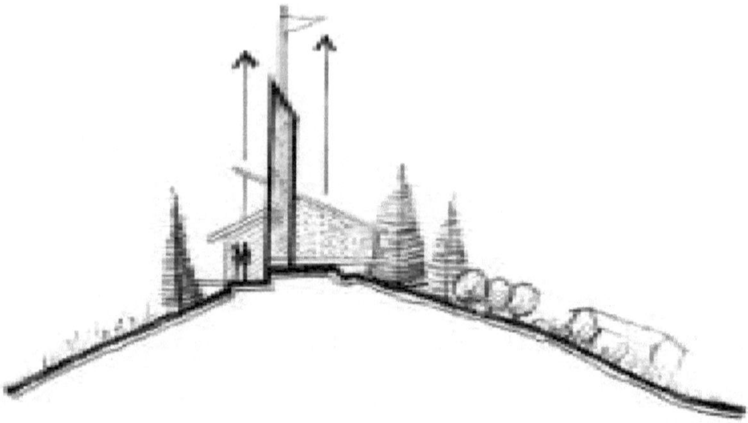

Elements located on convex landform are highlighted and become focal point

This type has no plain space. The slopes and crest of convex landform establish and perceive limits of a space and control views. It means steeper the slope, the definition of space becomes stronger. The convex landforms have aesthetic characters. It can serve as a focal point or dominant element (tall building) particularly when surrounded by lower, more neutral form like small building and trees. Another characteristic is that a person located on it will have general feeling of outward orientation. It provides superior viewing area and so it makes excellent building sight. It can also be used on North-West side of a space to block cold winter wind. It also lends itself to the dynamic and exciting use of falling water expressing movements.

3. *Ridge Landform:* A ridge is a high point of ground that is linear in its overall mass as compared convex landform that tends to be more compact and concentric. It is 'stretched out' version of convex landform. It defines edges of outdoor space and modifies microclimate on its slopes and in surrounding environment. It has general feeling of outward orientation. From a visual stand point, ridge has the ability to capture the eye and lead it along its length. Consequently, the ridge tops make logical locations

for roads, paths and other elements of circulation while the valleys in between are preserved an open space.

The other characteristic and use of ridge is a separator. As a special edge, the ridge acts as a wall dividing one space from another. It creates a feeling of "here" and "there". From a drainage point of view, a ridge functions as a "watershed divide". Water that falls on one side of the hill flows into one drainage basin or valley while water falls on opposite side flows into a different drainage basin or valley.

4. *Concave Landform:* It can be defined as a bowl like depression in the landscape. It is negativel solid and positively space. A concave landform can be created when earth is excavated from ground or when two convex landform are situated next to each other. This type of landform is an inward-oriented and self-centered space and focuses attention of anyone in space. The concave landform produces a feeling of privacy, isolation and protection from surrounding.

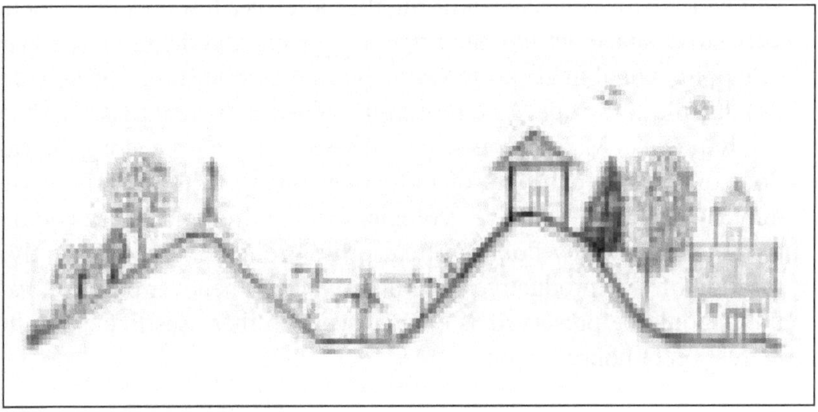

Concave landform provide a feeling of isolation and privacy

The enclosure and inward orientation of a concave landform make it ideal for making some enthusiastic creation which can also be observed from top of slope and steps can be provided to go down near the object. The concave landform has some other characteristics such as i) It is protected from direct exposure to wind, ii) It acts as a sun pocket where air temperature is higher due to direct exposure of slopes to sun, iii) It remains wet or water logged due to precipitation. Depending on area and depth, it can be used for lake or water reservoir or pond.

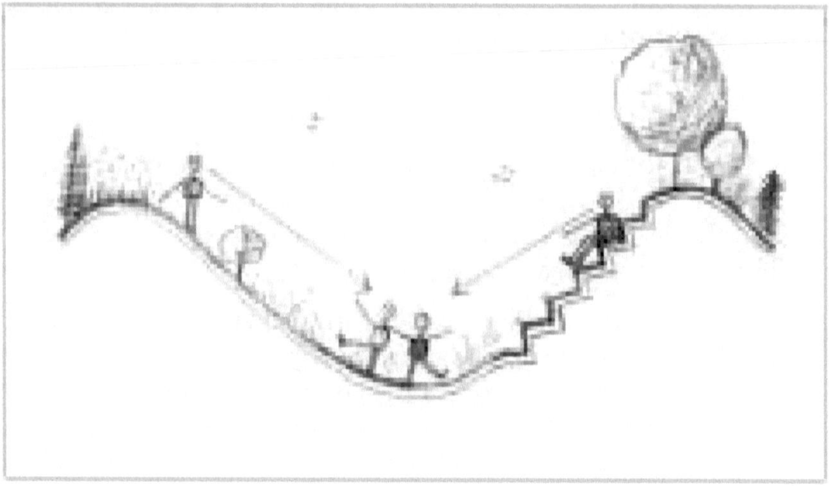

Conclave landform: Views are directed inward and downward

5. *Valley:* Like the concave landform, the valley is a low area that functions as positive space but like the ridge, it is linear and directional. Because of its directional quality, the valley is a suitable location for movement through the landscape. As for example valley of flower is a good location in Uttaranchal. Many forms of primitive travel occurred along the valley floor or streams or rivers of the valley owing to the relative ease of movement. It is a sensitive ecological and hydrological area compared to ridge. The valley floor is often fertile ground and consequently the location of very productive agricultural land. Whenever possible, valley floor should be preserved as open space for other uses like agriculture, recreation or conservation.

Functional Uses of Landforms

There are several functional uses of landforms in the outdoor environment. In all cases, the uses of landforms depend on the designer's skill and imagination. Landform may be used to create and define exterior space by excavation, filling earth and building up from existing base plane while convex landform is used by changing elevation to establish terraces. There are three variables of landform which critically influence feeling of space. i) The floor area of the space generally represents as usable area. Typically, larger the floor area, the space appears larger. ii) The steepness of enclosing slope affects our sense. These function as a wall of exterior space. The steeper the slope the more pronounced the delineation of space.iii) The horizon or silhouette line represents the edge between the perceived top of the landform and the landform and the sky. The space may seem to expand or contract depending on ones position in relation to horizon line.

Using variables of floor area, slope steepness and horizon or silhouette line, one can create limitless variety of spatial feeling. With constant floor area, three different spaces can be created by varying the slope and line of horizon. Likewise, the character of floor area itself may be changed to establish dissimilar qualities of space. One should keep in mind that slopes of 2:1 is quite sensitive to erosion and therefore it must be covered with ground cover and other plant material.

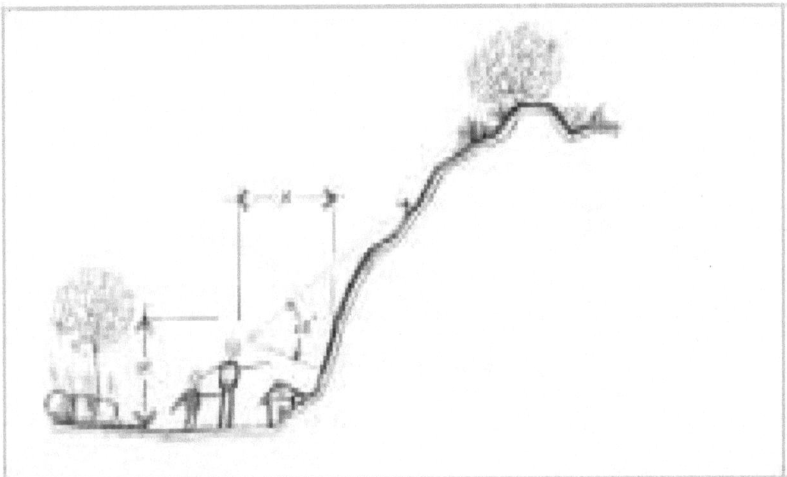

A landform provides a full spatial feeling when slope sides creates 45 cone of vision

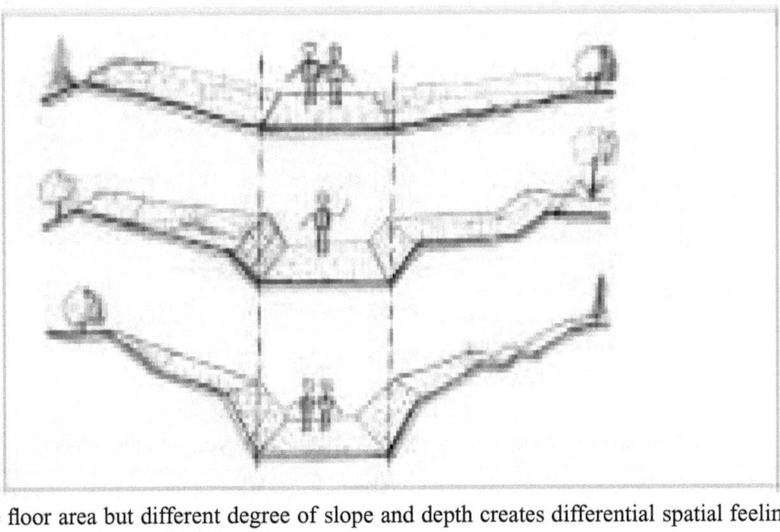

Same floor area but different degree of slope and depth creates differential spatial feeling like implied, defined and well defined space

Control view: Slope of landscape influence direct view to specific points in the landscape. It can create sequential viewing or progressive realization or completely block views from objectionable scenes. Landforms can be used to exhibit a particular object. Special sequences can be created that alternatively reveal and hide views of objects or scenes in direct or progressive way which develops a sense of anticipation and curiosity. One can hide an object at the slope which is not visible from long distance but as one come near the crest it is suddenly exposed.

Landform function as to direct view to a certain point

Landform partially hides particular objects directing view to other object

In a contrary way, landform may be built up in the form of earth mounds to screen out displeasing view or objects. Such techniques are used to hide roads, parking or service areas in a park by maintaining distance and height of view and viewer.

Influence movement: Landform influences the direction, speed and rhythm of both pedestrians and vehicular movement. As the slope of ground surface increases with obstacles then movements becomes more difficult due to use of more physical energy to get up and down. For walking, steps should be carefully placed otherwise the sense of balance is disturbed. So the slope should not exceed ten percent.

Affect microclimate: Landform can modify microclimate of area. South facing sloped surfaces can be used to receive direct sun in winter to get warmer surface and air temperature. Earth mounds or convex landform can be used to block cold winds while in summer these may be used to capture and channel the wind.

Aesthetic Uses of Landform

Landform can be used as a compositional and visual element. Earth is a material which can be pushed and pulled to create a desired form. Landform can be shaped into soft, sensitive forms which easily develop attraction while rock and concrete are used to form hard forms. Landform has an ability to be shaped in any way but it may also produce different visual effect under the influence of light and climate. Earth can be molded to form artistic structures or statues which may be placed in a garden as focal point or ornament. Landform serves as a background and a skeleton of a garden. Hence, its effect is dramatic and exciting when used accordingly. It has mystic effect when it hides certain objects and directs us to water fall or a water stream. The aesthetic effects of landform can be used with other elements to finally create a harmonious master piece of garden art.

Above all, the landform should be harmonized with overall appearance and other elements. The landform should not become too lumpy. It should have artistic curves. Landform can also serve as sole piece of architecture as a ground element for the building.

II Plant Material

Plant materials are extremely important design element and serve as the body of landscape gardening. It helps in combining and harmonizing different elements which also provide touch of life and beauty in an environment. Plant material, represents native and cultivated plants of all types, from ground cover to trees. It has potential of decorating a barren and harsh land. When designing the plant material one need passion, skill and whole knowledge about their cultivation.

Most significant characteristic of plant material is that they are living, growing elements. They are dynamic and so constantly changing size, colour, texture, appearance and overall character with growth and season. They require particular climate for survival and growth with proper maintenance. Plants provide a feeling of the live nature within environment especially in urban areas. Besides, aesthetic role, plants serve different functions which can be utilized in landscape gardening.

Functional Influence of Plant Material

Plant material are the only source of environmental moderation and serve as cleansing our air. Plant materials cleanse the air, retain moisture in the soil, prevent erosion and loss of soil and provide habitat for birds and animals. Further, plants create space or outdoor rooms, block unsightly views, stabilize steep slope, direct the movement through the landscape, visually unify a group of building and modify exposure to sun and wind. Plant materials contribute to real estate value of buildings and their sites.

Architectural Influence of Plant Material

Plant material may be architecturally used in the landscape to function as structural components like floors, ceilings, walls, windows and doors. Some architectural uses are given below:

i) Plant material can create sense of space by modifying the ground plane individually or in combination. One has to choose plant according to purpose after careful study of their overall characteristics. Some basic functional spaces created by plants are; a) Open space can be created by using low shrubs and ground covers. It is open in all direction. b) Semi open space can be created by enclosing one side with taller plants which act as a wall and other side planted with small plants. c) Canopied space can be obtained by growing mass of shade trees with a dense canopy. This encloses overhead space. d) Enclosed canopied space has same characteristics as canopied except that both sides are blocked with growing medium-tall to tall shrubs or small trees which provide privacy. e) Vertical space can be created by using tall, narrow plants which are vertical in orientation and open to sky. It gives upward movement. Other than individual spaces, Plant materials can be used to establish interlinked sequences of spaces. Vegetation may subdivide larger spaces created by two buildings into smaller spaces.

ii) Plants may be used to complete spatial definition and organization that has been suggested by design elements. Two common methods are as:

a) *Closure*: Plant material may be used to complete a defined space. A space surrounded by building or wall on two or three sides then vegetation can be used as a closure.

b) *Linkage:* Plant material in linear manner between the isolated elements visually type them together provide spatial enclosure.

c) *Screening:* Plant materials are used to screen unattractive or unwanted objects. Depending on objectives, a vegetative screen may be completely opaque to totally screen a view or it may have some of transparency to provide partial screening.

d) *Privacy control:* It is the technique of encircling a well defined area with plants of such a height that view from outside is prevented. The purpose is to isolate the space from its surroundings and for recreational activities for the family.

Aesthetic Influence of Plant Material

Plant material can also fulfill a number of aesthetic uses. Aesthetic uses of plant material can be in different ways like i) They are used as aesthetic decorative beauty with various colours, shape and structure of flowers and foliage. ii) They are used as specimen plant in garden. iii) They act as complementors as they complete design and furnish a sense of unity by repeating forms and masses of building or extending lines of building. iv) Plant materials are used as unifiers as they can visually tie together all different elements. v) Plants function is to emphasize certain points in the exterior environment. They do this by means of different size, form, colour and texture. vi) They also work as acknowledgers. It increases the value of location or site by making the space more obvious and easily recognized. vii) They may be used as a softener to soften the harshness and rigidity of architectonic shapes and forms. viii) Plant materials have direct influences on what is seen and the sequence in which the views are revealed as pointed out in the section on architectural uses of plant material.

Visual Characteristics of Plant Material

Visual plant characteristics like size, form, colour and texture affect the aesthetic quality of a design, all visual as given below;

Plant Size

Plant size directly affects the scale and proportion in a garden.

1. *Large trees:* Height of large trees is around 15-22 m. Eg. *Adansonia digitata, Anthocephalus cadamba, Azadirachta indica, Alstonia scholaris, Ficus sp., Casuarina equisetifolia, Couroupita guinensis, Peltophorum, Polylthia longifolia, Samanea saman, sterculia foetida, Terminalia cattapa, Tubebuia rosea, T. arjuna, etc.* These are dominant visual elements in garden. These are also used as focal point and for screening cold wind and hot sun.

2. *Intermediate Trees:* Intermediate trees have 12-15 m height. These are used for screening cold wind and hot sun. They are also used to develop enclosed area. Some examples are *Butea monosperma, Calophyllum inophyllum, Delonix regia, Kigellia pinnata, Jacaranda, Mimusops elengi, Markhemia, Tecoma argentia, Spathodia campanulata, Thuja actinophylla, Thespesia populnea, Bauhinia purpurea, Cassia fistula, Cassia nodosa, C. renigera, C. siamea etc.*

3. *Small trees and Ornamentals:* Height of small tree is 5-8 m. Eg. *Brassica actinophylla, Cochlospermum gossypium, Calophyllum inophyllum, Plumeria alba, P. rubra, Cytharexilon, Erythrina blackeii, Gliricidia, Largestroemia indica, Nyctanthus arbortritis, Saraca indica, Tubebuia avalandi*, etc. They are useful in creating edged for complete enclosure and also work as ceiling. They can also be also planted on both sides of path.

4. *Shrubs:* Shrub height varies from 1 to 4.5 m. Shrubs differ from small trees as being devoid of and having over all specific canopies but leaf mass. Shrubs allow vertical space with view towards sky. They also function to direct views to desired point, via path, help to screen unwanted scene, serve a background for any sculpture and as shrubbery. Eg. Hibiscus, Ixora, Rose, *Murraya exotica* (*Kamini*), *Tabernaemontana* (*Tager*), Lantana, Duranta, Acalypha, Eranthemum, Caesalpinia, *Cassia biflora, C. glauca,* Day-king, Night-queen, etc.

5. *Climbers*: Climbers play an important role in beautification of the garden with their special character to train on wall and other modified structures. They are grouped as follows: (a) Showy flowering climbers, *Adenocalyma alliaceum, A. calycana, Allamanda cathartica, Antigonan leptopus, Anemopaegama, Aristolochia, Argyreia, Bougainvillea, Tecoma redicans, Clerodendron, Clitoria ternatea, Tecoma grandiflora, Thunbergia alata, T. grandiflora, Petrea volubilis, Rose* (Climber

type), *Strophanthus*, *Stigmophylon*, Senecio, (b) Climbers with scented flowers, *Clematis*, *Echites caryophyllata, Jasminum grandiflora, J. Officinarum, J. Angustifolia, J. Nitidum, J. Sessiliflorum, J. auriculatum, J. sambac, Heptage medablota, Quisqualis indica, Lonicera japonica, Valleris hyenei, Passiflora* spp. etc (c) Shade loving Climbers with attractive foliage *Aspargus racemosa,* Asparagus plumosus, *Ficus repens, Hedera helix, Monstera deliciosa, Vernonia heynei, Scindapsus, Phillodendron, Syngonium,* and (d) Shade-loving flowering climbers like *Gloriosa superba, Ipomoea* spp., *Jacquemontia,* etc.

6. *Herbaceous perennial:* They are non woody plant materials generally used to make flower beds, edges and borders in general and also as pot plants. Eg. Canna, Heliconia, Amaryllis, Michalmas daisy, Perennial verbena, Pedilanthus, Perennial Chrysanthemum, *Vinca rosea*, Gerbera, Golden rod etc.
7. *Annuals:* Annual plants are used for making flower beds, edges, pot culture, hanging baskets, etc. They are used for display in different ways but need replacement in every season or year. Eg. Marigold, Gaillardia, Gomprena, Candytuft, Sweet sultan, Sweet william, Verbena, Coreopsis, Petunia, Salvia, Sunflower (ornamental), etc.
8. *Ground Cover:* They are low growing and spreading type. They are used to form spatial edge as lawn. They are used as carpet bedding. They enhance greenery and thus more charm, liveness and form a line between lawn and ground cover which create space and direction. Examples: Alternenthera, Setcreasea, Lawn, Chlorophytum, Pilea, Portulaca, Aerva, Eupatorium, Asysteasia, Wedelia, *Ternera ulmifolia* 'Elegans', Zebrina, Aptinia, Epesia, etc.

Plant Form

The form of an individual plant or group of plants is the overall shape and growth habit. Each form type has its own unique characteristics and accordingly can be used in garden as given below.

1. *Fastigiate or Oval:* This type is upright, narrow and tapers to a point atits top. Example: Thuja, Middle aged Eucalyptus and Casuarina.
2. *Columnar:* This is same as fastigiate form but has rounded top. Example: *Casurina equisetifolia, Polyalthia longifolia, Eucalyptus* sp., etc.
3. *Spreading:* This plant form is at least as broad as it is tall, with generally horizontal habit. Examples: All the palms.

4. *Round:* This form has rounded or spherical canopy. Example: *Mimusops elengi, Azadirchta indica,* Mango, *Anthocephalus cadamba*, etc.
5. *Pyramidal or conical:* It has cone like appearance. Example: Christmas tree, Deodar, Juniperus, etc.
6. *Weeping:* A weeping plant form has predominantly pendulous or downward arching branches. Example: *Celistemon chinensis, Weeping willow.*
7. *Picturesque:* It may be irregular, windblown and has unique sculptural shape. Example: *Butea monosperma.*
8. *Umbrella:* The canopy has umbrella shape. Example: *Samanea saman, Albezzia lebbeck, Gulmohar, etc.Shrub: Cassia biflora, Calliandra, etc*

Plant Colour

Plant colour is the most notable visual characteristic of plant material. It is emotional characteristic of plant material because it affects feeling and mood. Plant colour should be used to give effect to plant size and form in design. For instance, a plant used as focal point because of its size or form might also have a colour that further attracts attention. Darker green colour can provide a sense of solidity and weight with quiet, peaceful feelings while light green colour can provide airy quality to a space. It also gives feeling of cheerfulness and excitement in addition to moving away visually from the viewer. Colour attributes can be seen in foliages, flowers, branches and trunks.

Growth Pattern

It includes the form of the foliage and type of the growth and can be classified as:

1. *Evergreen plants:* Leaves remain on plant throughout the year. These are grouped at various locations throughout the design which visually break up composition and if clustered together they unify composition. Evergreen dark foliage is also used as background for light green foliage or for deciduous plants. They are used to screen objectionable views, cold wind and hot sun.
2. *Deciduous plants:* These plants lose their leaves in late winter, summer and regain in monsoon. This emphasis the season as they have different appearance in all seasons. These can be utilized to create mobility in garden. Deciduous trees have typical structure of branches. The habit and pattern of bare branches themselves create a unique design. Generally, type of foliage influences seasonal interest, visibility and unity of a design. They also relate directly to the texture.

Plant Texture

It is the visual roughness and smoothness of plant. It is influenced by leaf size, twig and bark size, bark configuration, the overall habit of growth and the distance at which the plant is viewed. Texture affects a number of factors in a planting composition, including unity and variety, perception of distance, colour tone, visual interest and mood of a design. Different texture displayed in plants is given below:

i) *Coarse texture:* Coarse texture is created by large leaves, thick, massive branches and loose open habit of growth. Examples are *Kigellia pinnata, Butea monosperma, Canophyllum inophyllum, Madhuca indica, Tectona grandis, Anthocephalus cadamba, etc.* It is highly visible, bold and aggressive so used as a focal point. Because of its strength, it causes the sensation of "moving toward" and make as outer space feelling smaller.

ii) *Medium texture:* Medium texture results from medium sized leaves, branches and moderately dense habit of growth. They are less transparent and form basic texture of a design, serving transitional element between coarse and fine textures. Example: *Spathodia campenulata, Tecoma argentia, Saraca indica, Mimusaps elengi,Cassia fistula*, etc.

iii) *Fine texture:* It is produced by many small leaves, tiny, thin branches and twigs and a tight, dense growth habit. They are soft and delicate in appearance. It increases distances due to its structure and also act as a neutral background to provide refined, smooth surface character. Example: *Cassia seamea, Delonix regia, Jacaranda,* Pink cassia, *Peltophorum, Casuarina, Samania saman*, etc.

III Water

Water is an extremely varied element whose characters and appearance depends on its external factors. Water may be used as purely aesthetic element with utilitarian functions. Water has a number of unique qualities like it has capacity to magnetizing and compelling all elements used in design. Humans are drawn towards water for both utility and visual purpose. Water is a basic need to support life not only physically but emotionally also. Water is considered to be a sign of purity and piousness and it provides a feeling of serenity and peace to the viewer besides imparting a touch of softness to the landscape. They are also an inevitable component for providing mobility to the landscape by means of the flora and fauna in and around them. Water, the most essential substance for the survival of all life forms, has also been given an important place in all the styles of landscaping *viz.,* Pertian, Mughal, Japanese, French, Spanish, Roman and Italian. Various water features have been used as garden adornments.

General Properties of Water

General properties of water are important to understand as these are very well utilized in landscape gardening. The important properties of water include:

i) *Plasticity:* Water has no shape and so determined by the characteristics of its container. Its particular form and appearance in any situation is the direct result of the influence of gravity.

ii) *Motion:* It is further classified into two categories according to its motion. S*tatic water* found in lakes, ponds, pools or gently flowing rivers. It is peaceful and relaxing in character with a soothing effect. It encourages thinking without interruption and has balance and equilibrium with gravity. *Dynamic water* is moving, flowing or falling water observed in rivers, streams, water falls, fountains etc. It is energetic and stimulating which easily capture the attention. It becomes exciting and dramatic on interaction with colour and light. Greater imbalance with gravity, with faster movement of water is observed.

iii) *Sound:* Water emits sound when it is in motion which influence human emotions. Example: the ceaseless, rhythmic motion waves against a shoreline may be quite and peaceful while roar of waterfall may be motivating.

iv) *Reflectivity:* Water has ability to reflect in its environmental setting. In quite, static state, water can function as a mirror, reflecting an image of its surroundings like land, vegetation, sky, etc. The reflectivity is affected by slope, size and shape and roughness of container, temperature, wind and light.

Functional Influence of Water

Water serve as different functional uses like

i) A common use of water is to irrigate lawn, gardens, parks and field crops etc.

ii) Water is used to modify air and temperature of ground surface.

iii) It may be used in outdoor spaces as a sound buffer to cut down noise where noise level is high. The falling or moving water can create a mask between noise area and viewer.

iv) Recreation is most important use of water in landscape as it is used primarily for Skiing, Swimming, Tubing, Windsurfing etc and secondarily for aesthetic use, boating, canoeing, kayaking, rafting etc.

Visual Influence of Water

Water is used with different garden features and design elements to develop an aesthetic value. Some of common visual elements are described below:

Flat/Static Water

Water can be used as flat, quiet water body in the form of pool and pond. Pool is a hard, well defined constructed container of any size filled with water. It may be of any geometrical shapes like square, circle, oblong or rectangle. Pools are used as a plane of reflection for sky and nearby object such as sculpture, trees, etc. It provides dark value or light value which depends on sky condition, the surface of the container and the location of the viewer. Static pool of water can be used as neutral background and foreground for other elements and focal point. It can be used as focal point itself as fish pool with lotuses and water lilies.

Pond is similar to pool except its design which looks more natural due to its soft, curvilinear forms. The sides of pond act as special edges affecting perception and views. Besides, accomplishing all functions as pools, a pond may create a feeling of repose and tranquility because of its soft, peaceful forms. A pond may be used in the landscape to establish a unifying link between different areas of garden. A sense of mystery and fascination is created when a portion of pond is seen disappearing behind hill or group of trees.

Flowing Water

It is any moving water confined to a well defined channel. Flowing water results when the channel and its bottom are slopped, allowing the water to move in response to gravity. It is best used to express movement, direction and energy. A more turbulent effect of flowing water can be created by developing wide and narrow bottom which is composed of rough materials to strike water and create sound.

Falling Water

Falling water occurs when water moves up and suddenly fall from elevation in form of channel. It expresses force of gravity. There are three basic types of falling water:

a) *Free-fall:* This type of water drops directly from one elevation to another in uninterrupted manner. Its characteristics are dependent on volume, velocity, height of fall and edge conditions. For instance, sound of splashing is different when water falls into water while sound of splashing gets intensified when it falls on a hard surface. Free-fall water can be used in form of waterfall as well as water wall (wall created by falling water).

b) *Obstructed fall:* This type of fall is caused by striking the water on various obstacles while dropping between two elevations. It produces more sound and thus easily noticed than free falling water.

c) *Sloped fall:* In this type, water is dropping along and down a steeply sloped surface. This is just like flowing water but occurs on a steeper slope in smaller controlled volumes.

Jets

Jet is another form of water display which is created by forcing pressurized water up into the air through a nozzle in defiance of gravity. Mostly it is used as a focal point due to its attractive up and down motion along with echoes. There are mainly five type of jets:

i) *Single-Orifice:* this is the simplest type of fountain jet, with water forced through a single opening nozzle and produces clear stream of water that is simple but striking in appearance.

ii) *Spray type:* A spray fountain jet is produced by many fine, mist-like streams of water that results from water being forced through a nozzle with many small openings. It is light and airy in appearance with soft "hiss" type sound.

iii) *Aerated type:* This type also has single opening but it has quite larger nozzle which produces turbulent sparkling white water effect.

iv) *Formed type:* A formed fountain jet is any type of jet that is shaped to furnish a special effect. "Morning Glory" and "Mushroom" are two commonly used jets.

v) *Compound jets:* A compound fountain jet is made by using several single nozzles in various shapes to give uniqueness in fountain. Sometimes combination of above mentioned jets are used to develop compound jets.

Visual influence and functional influence of water with their properties are used as an interesting feature in garden that emphasize the garden or exclusively as water park or water garden.

Water Park

A water park is an amusement park that features waterplay areas, such as water slides, splash pads, spray grounds (water playgrounds), lazy rivers, or other recreational bathing, swimming, and bare footing environments. Waterparks in more current states of development may also be equipped with some type of artificial surfing or body boarding environment such as a wave pool or a Flow Rider. Usually waterparks have different features like *Streaming Pool i.e.* Pool that streams water that is usually operated with tubes, *Wave Pool-* Pool that generates artificial waves, *Water Slide-* Slide with water, *Tube Slide-* Similar to water slide, *Boomerang Go-* it is extreme attraction with large tube. It pushes tube to the cliff zone and returns like boomerang and *Spas*-these are for fitness facilitating saunas and hot springs.

Water Gardens

Water gardens, also known as aquatic gardens, backyard ponds and garden ponds, have become popular in recent years. They have also been famous in Chinese and European history. Usually referring to a man-made feature, these gardens typically combine a pool with aquatic plants and often ornamental fish. Fixed items such as rocks, fountains, statuary, waterfalls and watercourses can be combined with the pool to add visual interest and integration with the local landscape and environment. Ponds natural or artificially created add extra charm and elegance to the garden. A garden created near a lake or a flowing river look soothing and appealing. Although, aquatic plants play important role in waterscaping. Aquatic plants add aesthetic beauty to the surrounding environment besides contributing to the environmental conservation. Many aquatic plant species are used as aquarium plants for indoor decoration. Many of them have medicinal value e.g. Sweet flag, Foxnut, Lotus, etc. Lotus is used for worshiping in the Indian mythology. Dried lotus pods contribute one of the major shares in exports of dry flowers and flower parts from India. It is popularly used as dry decoration in flower arrangements and framed articles. Lotus stems are also used as food in various dishes. Water lily can be grown in small pots or tubs in home gardens. Iris and Calla lily are used as cut flowers. The essential oil extracted from the roots of *Acorus calamus* and Orris are highly priced for their medicinal and aromatic respectively. Aquatic plants can also be used for pollution control in water bodies. Effective reduction in heavy metal content in the water bodies can be achieved using these aquatic plants which serve as an economical and eco-friendly way of pollution control. Azolla, besides being an important biofertilizer, is also efficient in removal of mercury effluents form water sources. Aquatic plant species used as fodder for livestock include *Ipomoea aquatica, Echinocloa* spp., *Salvinia, Oryza* etc. They are used to conserve and beautify marginal lands and attract birds and butterflies. *Salicornia* is a genus of succulent, salt tolerant plants that can grow in salt marshes. Besides *S.europaea* and *S.brachiata* are edible, the seeds of *S.bigelovii* contain high level of unsaturated oil and protein and have immense potential for production of biodiesel. Lotus, Foxnut and Chestnut are popularly used as food. The starchy white seeds of *Euryale ferox* known as foxnut (*makhana*) are edible and are used widely especially for making a variety of delicious dishes etc. It also has medicinal and nutritional value and is very good for patients suffering from anaemia. The aquatic plants are basically classified as submerged and oxygenators, floating aquatics, surface flowering and as marginals.

Submerged plants and Oxygenators-Aquatics with foliage underwater are called submerged water plants. They add a larger amount of oxygen to the water and also help to remove nutrients from the water, which help to control algal growth. Most submerged plants thrive in water depths of upto 10 feet, but some varieties need a little light and grow at depths of 15 to 30 feet. Too many submerged plants can cause wide pH swings so it is important to maintain their population. Some of the important submerged/oxygenators are: *Elodea canadensis, Hottonia palustris, Ceratophyllum demersum, Sagittaria natans, Vallisinaria americana* etc. These plants are widely used as aquarium plants for indoor decoration.

Floating plants-Floating plants have hair-like roots that protrude into the water, need no soil at all and keep water clean by absorbing nutrients. They provide shade to the fish from the hot sun and protect them from predators. Water Hyacinth, *Azolla filiculoides, Lemna minor* (Duck Weed), *Pistia stratiotes*, Water Chestnut, *Euryale ferox (foxnut/makhana)* are typical examples of floating garden pond plants. *Trapa natans* (water chestnut/*shingara*) can be eaten raw or boiled. When the fruits are dried, they are ground to make flour which is used in many religious rituals and is consumed as a *Phalahar* on the day of fasts. Water Hyacinth i.e. *Eichhornia crassipes* is a very important plant for phyto remediation but has become a weed.

Surface flowering plants-Plants having their roots submerged in soil with leaves floating on water and flowering at the water surface are referred as surface flowering aquatic plants. Lotus and water lilies are beautiful example of surface flowering aquatics. Lotus, the national flower of India, symbolizes victory, survival, divinity, fertility and purity. Lotus is used for worshiping Gods and Goddesses since ancient times in the Indian mythology. Lotuses are among the most ancient plants. Their viable seeds have been found dating back to more than 2000 years. There are two species within the Lotus genus, *Nelumbo- N. nucifera* and *N. lutea.* Lotuses are sun loving plants and need at least 6 hours of sun light, daily. They have a waxy coating on their leaves. Lotus cultivars grow from 6 inches to 6 feet in height. Smaller selections, called Bowl lotuses, can be easily grown in containers less than a foot wide, in just a few inches of soil and water. Lotus cultivars are available in wide range of sizes and colours. Fragrance varies from fruity to mild. Colours range from deepest rosy pink to cleanest white. There are also bicolours and tones that blush or fade. Lotuses can be propagated by seeds as well as by divisions of rhizomes. Most water lilies grow in 3-4 foot depths, some in depths of upto 8 feet, but they do quite well in ponds with only 6-18 inches of water over their roots. Water lilies are classified into two broad groups— hardy water lilies and tropical water lilies. Hardy water lilies survive winters in cold climates. Tropical Water Lilies are day blooming cultivars and night bloomers.

Marginals- These grow at the edge or margin of the pond where the soil is moist or the water is shallow. They are well suited as transitional plants that link the pond with the other garden because of their adaptability to grow in a diverse range- - from wet/moist soil to submerged soil (few inches to 2 feet). The main aim of planting marginals is to decorate the pond side/landscape and adding colour, texture and form to the design. Some of the important marginals are- *Acorus calamus* (sweet flag), *Caltha palustris* (marsh marigold), Water canna, *Carex* spp.(Sedge), *Colocasia esculenta* (Taro), *Cyperus* spp., *Dichromena colorata* (white-top sedge), *Hibiscus moscheutos, Houttuynia cordata 'Chameleon', Iris* spp., *Ludwigia* spp., *Marsilea* spp., *Mentha* spp., and *Typha* spp. *Colocasia esculenta* has heart shaped leaves. It is a perennial having high to medium water requirement. *C. esculenta* var. *Black Magic* has black to dusty purple foliage which can be used with any silvery foliage plants along the edging of a water garden.

I Landform

Level landform Convex Landform

Ridges

Valley

Valley

Plate 17

II Plant Material

Acroclinum

Altea rosea

Antirrhinum majus

Annual sunflower

Arctotis

Californian Poppy

Briza

Plate 18

Bromus

Calendula

Candytuft

Chrysanthemum coranarium

Coreopsis

Dahlia

Dianthus barbatus

Dimorphotheca

Euonymus fortuneii

Plate 19

Gamelopis

Gazenia splendens

Geranium

Helichrysum

Hyacinth

Lagurus

Limonium

Melucella Leavis

Marigold

Plate 20

Mesembranthemum

Nasturtium

Nigella

Ornamental cabbage

Pansy

Pimpinellifolium

Poinsettia

Shirley Poppy

Plate 21

Primula malacoides

Ranunculus

Salvia

Sweet Allysum

Sweet Willium

Plate 22

Bulbous plants

Agapanthus

Amaryllis

Clivia miniata

Day lily

Deffodil

Spider lily

Lilium

Plate 23

Cactus and Succulents

Aeonium arboreum

Agave schibigera

Begonia rex

Dizygotheca elegagantissima

Echeveria peacockii

Euphorbia Thai hybrid

Fuchsia sp

Furcraea

Plate 24

Gynura aurantiaca

Hawothia fasciata

Hedera helix

Hyacinthus

Kalanchoe tomentos

Lampranthus roseus

Neoregelie

Opuntia microdasys

Pereskia acculata

Portulacaria

Plate 25

Cycus

Cycas

Zamia furfuracea

Ferns

Nephralepis exelata

Nephrolepsis biserrata

Sedum morganianum

Sedum rubrotinctum

Senecio rowleanus

Senecio serpens

Plate 26

Tradescantia sillamontana

Trees

Adansonia digitata

Alstonia scholaris

Amhrestia nobilis

Araucaria

Butea monosperma

Bauhinia pupurea

Beech tree

Plate 27

Callistphus chinensis

Calophyllum inophyllum

Cassia fistula

Cassia nodosa

Cassia renigera

Couropita guinensis

Cochlospermum gossypium

Delonix regia

Erythrina crista-galli

Plate 28

Jacaranda

Schleichera oleosa

Magnolia

Michekia chamaca

Mimusops elengi

Nerium oleander

Peltoforum

Pisonia alba

Plumaria

Plate 29

Plumera

Polyalthia longifolia

Prunus

Pterospermum acerifolium

Salix babylonica

Samanea saman

Sarrica indica

Spathodea campanulata

Tabebuia

Plate 30

Tabebuia heterophylla

Acalypha

Adenium

Aglonema

Allamanda

Alocacia

Aralia

Bougainvillea

Canna

Plate 31

Clerodendron thomsonae

Croton

Dombeya

Dracaena

Eranthemum

Euphorbia leucosephala

Excoecaria bicolour

Hamelia Patens

Plate 32

Hibiscus

Ixora

Lantana camera

Maranta

Mussaenda

Nerium

Pachystachys

Phyllanthus nivosus

Plate 33

Plumbago ovata

Poinsettia

Schefflera

Syngonium

Tabernaemontana

Tecoma stans

Plate 34

Climbers

Adenocallyma allyaceanum

Antigonon leptopus

Thunbergia grandiglora

Vernonia

Wisteria chinensis

Combratum comosum

Quisqualis indica

Jaquomontia violacea

Peparomia

Plate 35

Petrea volubilis

Stigmophyllon ciliatum

Hiptage benghalensis

Tecoma grandiflora

Thunbergia alata

Herbaceous Perennial

Begonia rex

Geranium

Plate 36

Periwinkle

Mirabilis jalapa

Coleus

Kalanchoe

Maichalman Daisy

Plate 37

III Water

Flowing

Stastic

Plate 38

CHAPTER 6

Components of Landscape Gardening

Definition and Significance

Garden Components are the garden features that form the expressions of garden, adding charm to the basic garden landscape and constitute its style, contributing strongly to its texture and structure. Thus, garden features are important as they provide face lift to the garden skeleton, giving it a filling effect with charismatic appeal. Garden features include garden wall, fencing, arches, pergolas, hedge-edge, topiary, etc. Creating sudden and unexpected change of scenery sometimes reveal good contrasts. Thoughtful planning of garden features make an invaluable contribution to garden design. It is possible to design a garden with a large number of man-made artistic items. However, indiscriminate and haphazard use of these items results in clutter. Further, new and strange features in the garden always attract visitors and give unbounded pleasure. During ancient time, because of the traditional style of architecture and high craftsmanship, garden features and adornments have played a big role in many famous gardens of the world. Selection of features and positioning of ornaments in a garden are the visible evidence of the owner's taste of nature and aesthetic sense. Whatever features are chosen, they must add charm, create an atmosphere of naturalistic design and improve the quality of landscape design. Garden features and ornaments are many, but they must be utilized in the design with restraints. Some important garden features are described below.

Garden Walls and Fences

Purpose: Garden wall serves to provide boundary to the garden, provides screening, privacy, safety, security and protection from outdoors, edit views, screen out wind and noise, hold the elevated soil structure, modify climate, modify visual elements, demark a garden section or sometimes even act as a hedge and may be a feature inside the garden. Besides, low free standing walls may also be used to serve as sitting element. To serve as comfortable sittting the wall height should be 18" and 12" width. Although, fence also provide the same function, but walls serve as more permanent structure. Thus wall and fences define space, provide hard, stable, well defined vertical planes.. High walls enclose space and low walls imply space to create outdoor rooms. Walls and fences may be used to totally blockviews or partially screen views in varying degrees depending on the material chosen and the height selected. Height of the wall or fence strongly influence the degree of privacy needed. Although, tall wall or fence is required for total privacy, a lower wall or fence is adequate for people sitting or lying down. Walls or fences are often used to separate seprate adjoin uses or function on the same plane like a quiet sitting area from a parking lot, a swimming pool from general public area, etc. Walls may provide additional function of minimizing negative effects of harsh Sun rays and wind through proper direction planning and material selection. As for example, a wall erected on the west or southwest side of a building can prevent heat build up during late afternoon time and reduce wind influence. When walls are erected to reduce wind influence, it is necessary to have some openings within the wall to provide better wind direction and avoid turbulence. Walls or fence may also imply visual effects, they may act as neutral background to other elements in the garden like plant material or piece of sculpture. Walls also help to unify building and the garden site and thus serves as linkage of different elements. Further, the design chosen for the wall, the colours or graphics selected and the mode of construction in straight lines or curvy, straight or in and out, influence visual interest in the viewers.

***Type of walls*:** Walls are either as freestanding or retaining. Free standing wall are self-supporting. Retaining walls hold back a slope or volume of earth from a lower elevation. Compared with slopes, retaining walls establish sharp distinct edes, distinct edges and planes that end to be more prominent from a visual standpoint. A garden should be such that, one should enjoy from inside and should also be visible from outside. But sometimes, from safety or beauty point of view (*e.g.* to obscure the ugly sight of an open drain), it may be necessary to erect a brick, concrete or masonry block or stone wall along the periphery of the garden. Sometimes mixed material wall made up of screen blocks with bricks and stone panel may also variety. Stone may cut or uncut for the wall. In its uncut state, field stone is generally use. The construction can be masonry or wet wall i.e by using cement or concrete block and mortar, or dry wall i.e where no binder or mortar is used (for more natural effect and only for low height). Bricks provide warm, refined quality and creates more smoother and polished surface. Bricks can be employed in number of potentiall methods in courses as a) stretechers b) headers) c) soldiers forming different patterns like running bond (more common),English bond and Flemish bond. Openings or spaces within the walls may provide some views through the wall as well as makes it less massive and heavy in appearance. A compromise is to have a low brick (or concrete or stone) wall of 60-90 cm height and to put over it some grills. However, tall solid walls (more then 6') are most effective in screening provision that is particularly required around parking lot, roadways, park or around parks or unsightly industrial equipment. Here, its important to avoid eye level or 6'height of wall as it doesn't completely block the outer view but creates disturbing views of the objects beyond. In case of construction of retaining wall on slope, it could be in three ways i) top slope is cut away and soil moved downhill, ii)soil is cut away from the mid portion below the wall and filled on the top and iii) slope is divided into two terraces with two walls and soil is filled from the bottom potion. Dry laid stone, timber cribbing, broken concrete, vertical timber boards or brick veneer on concrete blocks can be used for construction of retaining walls.

Regardless the type of wall is selected, it is necessary to have solid concrete footing twice the width of the wall as the strong foundation. To break the drabness of a concrete or brick wall, one may grow creepers such as *Ficus repens* over the wall. Walls can be substituted by a group of shrubs or trees also. Growing tall shrubs along the wall emphasize its liveness and if planted with colourful low herbaceous perennials add charm and show attractive contrast between artificial qualities of wall and naturalistic vegetation. Cavity walls i.e walls having deep crevices that can be lined with colourful plants may look highly attractive. Sometimes brick wall with integral seat and planter may also serve multipurpose use. These are also termed as "green live wall". It becomes an object of beauty. Plants growing in the crevices of the stones and hanging down the face of a dry wall look beautiful and are a common feature of the English gardens. Live walls are generally retention walls, with only one surface available for gardening or planting while the other side is supported against the vertical or near-vertical face of two different levels of landscape. A cavity dry wall can also be constructed as a column. It may be possible to convert a portion of the concrete wall or garage wall with very little additions into a live wall.

***Plant selection*:** The plants suitable for live wall are *Ageratum,* Sweet alyssum, *Gazania splendens, Gypsophila elegans, Mirabilis jalapa, Oxalis sp., Portulaca, Verbena* (perennial and annual), *Viola, Zebrina pendula (*Syn. *Tradescantia zebrine), Zinnia linearis etc.* Miniature roses and some other small herbaceous plants can also be grown on the top of the wall. Some ferns and other plants suggested which may be very useful to give natural effect to the wall are *Adiantum caudatum, A. venustum, Polystichum sp., Lycopodiums*, etc.

Elements of walls/fence: In the design of walls or fences, three important elements are a) base b) wall or fence surface and c) cap or top. The base need to be visually strong and well established. Base of wall that steps down slope creates visually stable support as compared to the one that just parallels the slope. Wall or fence surface depends on aesthetic, spatial, functional or budgetary objectives. Horizontal material pattern provides stretched effect while vertical lines provide tall and compact effect. Cap functions to cover the body of the wall or fence and prtotets it from the infilteration of water. The cap also provides finishing touch and complete look to the fence. A wider and visually heavier cap to the fence surface may be used for more visual emphasis.

Fence-Fences can be defined as free standing vertical planes constructed of either wood or metal and these are typically less thinner and less massive than walls. Fencing, besides marking the garden boundary, provide privacy, safety and protection. Sometimes, fences are also put to separate one part of the garden from the other. Further, fences are cheaper than the walls, involves less time in installation, offers wide range of material or feel, allows better view from outside although they are less durable than walls and less capable of reducing noise pollution from outside Several materials such as wood, bamboo, wrought iron, wire, wire nettings, and chain-link fences may be used for this purpose. Depending upon material that can withstand weather and years of constant use should be used in making fences like louver, split rail, alternating boards, pickets, board and lattice or grape stake. Wood and bamboo provide a range of visual effects from rustic and naturalistic to smooth and formal or informal depending on its finish. Fences of wood and bamboo get spoilt quicker than the wire fences and demands special care and maintenance of painting and staining. Wrought iron fences are more used in formal styles and looks elegant against a light background and dramatic when sunlight shines through them from the background. The Fence design may be solid like louver or clap board or open like picket, split rail, grape stake or wire fencing as per the degree of visibility. The simplest wire fencing will be to put stout wooden or concrete posts at 3 m distance and pass the wires through these posts starting from 22.5 cm intervals at ground level to 30 cm upwards up to height of 1.8 m. Wire netting fence is cheaper and keeps away animals. Chain-link fencings of various designs are available and these are fixed in angle iron poles spaced at 3 m distance. Wood fencing needs special care of painting and staining before installation, followed by painting at regular intervals thereafter. Since, wall and fence serve the same function, both the features may be combine like having wall upto 3′ h and having wire or wrought iron or wooden picket fencing upon it. This way, better visibility and openness may be provided along with strong protective measures.

Quick growing and elegant creepers such as *Bignonia venusta*, 'Honeysuckle' (*Lonicera periclynemum*), *Vernonia elaegnifolia*, railway creeper, etc. can be grown along the wire net fencing to make it more attractive and to reduce its openness. However, thick and strong creepers like Allamanda, *Tecoma grandiflora*, etc may be avoided as they tend to change the shape of the wire fencing.

Gates

A garden gate i.e the entry into landscape garden should be in harmony with fence and wall and should blend with the landscape. It should be inviting, stylish as well as functional. It also confirms the style statement of the garden like a low picket and airy lath gates looks people friendly while a high solid gate shields a private world within. The width of the gate is dependent upon the existing fence or wall size, however, a basic gate consist of a rectangular frame of 3-4′ by 8′. A high quality material that can withstand weather and years of constant use should be used in making gates like louver, alternating boards, wooden, bamboo, board and lattice etc. The material can also be mixed like board in the half way from bottom and airy and see-through criss cross canes above it makes it more interesting. The gate should have strong hinges and latches as well.

Trellis Screens

Trellis screens either free standing or along the garden wall are excellent a support for the climbers and as display of climber texture and flowers, besides offering an effective method of screening or dividing one part of garden from another in a beautiful manner. Generally wooden canes, bamboo or even poles can be used for making criss cross structure of trellis screens. The height could be around 8′ for screening purpose or lesser if placed along the wall. The criss-cross width could be around 6-8”. It is necessary to install the trellis strongly in the soil with some concrete work as has to support the strong climbers from falling or bending.

Raised Beds

Raised beds are beautiful attractions of a garden and their shape, style and material influence the landscape style and personal choice of the owner. On sloping sides, raised beds provide level areas for planting while on flat planes, they create needed visual relief, stimulate visual interest and elevate special plantings into positions of prominence. When placed against wall, it provides transition between vertical and horizontal lines in a beautiful and softer manner. They are extremely suitable for plants which thrive well in extremely well drained conditions. Providing more suitable and attractive setting for plants, as it allows drooping and crawling plants as well as dwarf plants to be viewed at close hand. They create and emphasize patterns in the landscape and provide beautiful site for displaying of colourful annuals and trailing plants. If provided with broader wall, may also serve seating. In heavy soil conditions, raised beds provide facility to grow climbers, vegetables, herbs and ornamentals that need reasonably light well aerated soil-media. Functional use of raised beds include control over soil composition and drainage, can even work well in case of poor soils as required compost can be filled in the raised beds, provides elevation to plants where working practices become easy and provides a clean and well maintained look in the garden. When placed against the garden wall, they act as transition between ground and wall and soften the hardscape. Raised bed offers more then one surface for growing as some plants love to trail downward while some may like to grow in the crevices of the raised bed wall. Well integrated raised beds in the garden enhance its overall look along with nicely laid paving provides low maintenance beautiful landscape. A raied bed near the building area also accentuates the building besides forming trastion between the garden and building. According to the sunlight availability the raised bed near the building could be provide a soft green look with indoor plants or bright look with annuals. A variety of material like bricks, concrete, wood logs, stone, cut stone, concrete blocks, boards etc can be used for raised beds. Also the size could be selected as per the personal choice and style demand, however 2-3' height is regular and preferred. Care should be taken to provide drainage with few holes at few inches from the bottom of the side walls. Also inside the raised bed at eh bottom 4-6 inch should be filled up with free draining material like brick pieces or thick gravel. A raised bed sorrounding a tree also looks elegant with dwarf heraceaous trailing plants besides providing protection to the root system as well seating area with wider walls. Annuals with dwarf habit and profuse flowering like mesembranthemum, nastrutiums, sweet alyssum etc add enthusiastic visual interest to the tree.

Hedges and Edges

Hedge: Shrubs or trees planted at regular intervals to form a continuous screen or form a live wall is called 'hedge'. Purpose and Usefulness of hedge are many. They act as a compound wall, gives shelter from strong grails, also ensures privacy, i.e. serve the purpose of a screen, forms a background for a floral display such as herbaceous border, separates out one component of a garden from the other (e.g., the vegetable garden from the flower garden or a lawn from rose bushes), helps in screening the ugly and unwanted spots such as manure pits, servant's quarters, etc., in the garden. A live and impregnable hedge, reinforced with barbed wire, is inexpensive compared to a boundary wall and if maintained properly looks better than a dead and dry boundary wall.

Criterion for selecting a hedge plant: In a garden hedge is planted with two objectives and accordingly plants are selected:

a) *Protection against theft, trespass, wind:* Plant should have the characteristics such as quick growing (with height 2 m or more), hardy, drought resistant, thorny, dense, responsive to frequent pruning and clipping. The plant should have the capacity to rise quickly by seeds or cutting. The plant height may be 2 m or above.

b) *Ornamental purposes or screening:* The plants have attractive foliage and/or flowers, should be dense in growth habit, and stand regular clipping. The ornamental hedges are generally low in height and do not obstruct the view completely since the other portion is visible over the hedge. The height may be 60 cm to 180 cm, depending upon privacy need.

Planting Process: An ideal hedge is that which makes uniform dense growth from base to top. Each plant in the entire length of the hedge should get more or less uniform supply of water, essential nutrients and sunshine. Next process is the preparation of the land. In a hedge the plants are spaced much closer than they would have been for any other purpose. Moreover, once a hedge is raised it is likely to remain there for about many years. Digging a trench 30-90 cm wide and 30-75 cm deep, depending upon the type of the hedge. The heavier the hedge the deeper and wider is the digging. Deep digging also encourages the plants to root deeply and prevent sending less number of laterals to compete with the plants growing along the hedge. The digging operation should start several weeks prior to actual planting time. A moderate dose of well rotten manure is mixed with the soil. If the manure is spread over the top layer, it may attract white-ants also. Under Indian conditions, the planting of hedge is undertaken at the beginning of the monsoon season in June-July. A hedge is started by sowing seeds or by planting cuttings *in situ* or by planting rooted cuttings, generally in double rows. Trees such as *Putranjiva roxburghii, Casuarina equisetifolia* and *Polyalthia longifolia* may be planted 75-90 cm apart. The spacing for shrubs will vary from 15-45 or 60 cm. The seeds are sown or cuttings are planted 15-60 cm apart in the double rows when the hedge is about 15-25 cm tall, it is topped back to 10 cm garden shears. Heading back should not be delayed to achieve quick growth as vertical growth occurs at the expense of basal growth. The lateral growth should also be topped in the same way. Weeding, watering, clipping and pruning are important operations to keep the hedges neat and in good shape.

Selection of suitable plants: There is a wide variety of trees, shrubs, and other plants which can be formed into hedges. A list is of plants for general guidance are a) Palms as Hedge: *Areca lutescens, Ptychosperma macarthurii, Rhapis excelsa, etc.*, b) Trees and Conifers as Hedge: *Acacia, Casuarina equisetifolia, Cryptomeria japonica, Erythina indica, Grevillea robusta, Inga dulcis, Parkinsonia aculeata, Polyalthia longifolia, Pongamia glabra, Putranjiva roxburghii, Thevetia* *nerifolia, Thuja, etc.*, c) Shrubs as Hedge: *Acalypha, Aralia, Barleria, Bauhinia acuminata, Bougainvillea, Caesalpinia pulcherrima, Carissa carandas, Clerodendron inerme, Croton, Daedalacanthus nervosus, Duranta, Eranthemum, Galphimia gracilis, Graptophylum, Hamelia patens, Hibiscus, Justicia, Lantana, Lawsonia alba, Malpighia coccigera, Murraya exotica, Plumbago capensis, Poinsettia pulcherrima, Rosa, Sesbania aegyptica, Tecoma stans, etc.* d) Cactus, Succulents and other Hedges: *Agave americana, Bambusta, Cereus, Euphorbia antiquorum, Furcraea selloa, Opuntia, Pandanus, Pedilanthus tithymaloides,* etc.

Edge: Lining of borders of flower beds, paths, lawns and shrubbery with brick, concrete, living dwarf plants, etc. is known as edging. Certain plants highly suitable for edging purposes are known as edge plants. Edge provides a lining only, for the purpose of decoration or demarcation. Edgings are formal or informal.

Formal Edging The materials used for formal edging include

i) *Bricks*: This is most commonly used for formal edging. The brick selected should be of good quality. The edging is to be put in a straight line or at a 45^0 angle to give a serrated finish.

ii) *Tiles*: Tiles of various shapes and sizes are available which are quite suitable for edging. It is much more difficult to keep the alignment of the tiles than that of bricks.

iii) *Stones*: Stone slabs used for edging are also easily available. Stone edging preferably should have rounded finish.

iv) *Concrete*: Concrete edging with rounded and attractive tops look quite attractive. Precast concrete edges in lengths of 1 to 1.5 m are available which require less maintenance.

Informal Live Edging: A strip of grass or low growing plants are an attractive edging especially in front of flower beds. A grass should not be less than 60 cm in width, and must be properly maintained. An informal edging of stone on crazy path with prostrate plants growing in between serves as a very good edging in front of a border.

The plants selected should be dwarf, bushy, hardy and should have foliage or flowers of a lasting nature.

Edging Plants Having Attractive Foliage: *Aerva sanguinlenta, Golden duranta, dwarf Eranthemum, Alternanthera, Asystasia coronandeliana, Cineraria maritime, Coleus, Echeveria, Eupatorium cannabinum, Iresine, Pilea muscosa,* etc.

Edging Plants for Flower purpose: *Alyssum, Amaryllis, Brachycome, Calendula, dwarf marigold (T.patula), Zinnia lineris, Gerbera, Torenia, Zephyranthes different types, Pansy and Perennial Verbena. Miniature roses can also be used for edging.*

Arches, Pergolas and Overheads

Arches

A garden may need arches for training climbers or ramblers and are generally constructed near the gate or over paths in the garden. An arch should be at least 2 to 2.5 m high, so that the branches of creepers hanging down should not interfere with access. The breadth should not be less than 1 m. Generally it is as wide as the gate. The basic material used for making arches is rustic poles, iron or sawn timber or bamboo. Galvanized wire netting may be fixed on the sides of the arches to help the creeper to climb up. Characteristics of climbers used for arches include light weighted, thornless, flowering type preferably with profuse and scented flowers. Evergreen arches may be erected as entry or exit gate at intervals of 45-60 cm. Climbers suitable for Arch are *Antigonan, J. auriculatum, J. sambac, Allamanda, Petrea volubilis, Combretum, Passion flower, Jacquimontia, Asparagus racemosa, A. plumosus, Tecoma grandiflora* etc

Pergolas

A pergola may be defined as a series of arches joined together. Pergolas are generally constructed over pathways, which add beauty to a garden. A pergola is a beautiful resting place during the summer months in a tropical country. The path below remains cool due to the creepers growing above. If the pergola is wide enough, concrete or wooden benches may be constructed for sitting. Like arches the support can be made of wooden or stone or brick pillars, angle iron and G.I. pipes. The support posts and other parts of pergola must be very firm, as they will have to withstand strong gusts of wind and bear a weight of the chosen climbers. The roof may be made of iron angles of different sections with longitudinal and cross rails. Over this base strong galvanized wire mesh or welded mesh may be placed for the creepers to spread easily. The width of pergolas is kept generally within 2-2.5 m and the height is also the same but preferably on the higher side. Characteristics of climbers used in pergola are:

i) Profusely growing flowering behavior,

ii) Flowering or only foliage type

iii) Should be evergreen to provide shade & coolness

iv) It should be thornless and should have dense foliage. In broad pergolas it may also be possible to keep a few shade loving plants.Climbers suitable for Pergola: *Allamanda, Adenocalyma, Clerodendron splendens, Quisqualis indica, Hiptage medablota, Thunbergia grandiflora,* Bougainvellea, Rose, *Ipomea palmate, Pyrostagea venusta, Stigmophylon, etc.*

Overheads/huts

Overhead screen provide shelter, protection and privacy, expand outdoor living space and help connect these spaces to other features of the garden. Further, overhead screens have functional use of having tea time, enjoying rain and landscape sitting silently, adding immense pleasure and outdoor enjoyment. Further, along with screens overheads provide protection from sunny rays and wind. Overheads look extremely beautiful when climbers are supported over it by creating arbour. A variety of material like wood, canes, bamboo, poles, boards etc can be used to make overhead frame or skeleton. Further, same frame can be roofed using glass, wood lath, lumber, reed mats, acrylic, glass panels, canvas or shade cloth as per the need and budget. The solid roof should be slightly pitched and height should be around 8′. Climber suitable for pergola are also suitable for overheads. With a more solid structure even outdoor room type of structure may be created like gazebo, garden pavilion, patio room, or even garden work center.

Huts in a garden is unusual feature which create an interesting soothing effect on the visitors. It is a place to seat in shade in garden and can be used as place for family to have tea or checks in the garden. These are made of concrete or bricks walls. Huts made of straw give natural look and are thus mostly preferred in garden restaurants. Benches and seats with tables should preferably be arranged in the garden hut.

Lawn

Lawn is a basic feature for home ground development and an essential feature for any type of garden. It can be defined as the green carpet for a landscape. It improves the appearance of the house, enhances its beauty, increases, convenience and usefulness, thus adding monetary value to the real estate in home garden. Further, the reen stretch of lawn provides perfect setting for a flower bed, border, shrubbery or a specimen tree or a shrub. It has also spiritual value as it manifests stability, peace and calmness and is an important source of charm and pride.

The view of a lawn should be uninterrupted from the house or from the entrance. If the lawn is spacious, seasonal beds, herbaceous borders, beds of canna, attractive specimen trees like *Cassia renigera, C. fistula,* Bottle brush,

Spathodia, Plumeria alba (evergreen), *P obtuse, Tecoma argentia, etc* or shrubs such as *Araucaria cookie, Thuja, Callistemon lanceolatus, Brya ebenus*, etc., may be planted. Some bulbous plants such as *Zephyranthes, day lily, Amaryllis, Hippeastrum* may also be planted in groups in the lawn. Grass is one of the hardiest perennial herbs and it is very difficult to maintain a lawn, although one has to take the necessary care.

Site and Soil Type

The site for lawn establishment should receive full sunshine. Best situation will be southern side and next best is the south-east and south-west of the building. No big trees should be present at the site because the falling leaves make the lawn dirty. Soil moisture and drainage should be optimum. However, there are some shade tolerant lawn which can be developed in shady locations. A fertile loamy soil containing enough organic matter is best suited for lawn preparation. Depth of the soil should be 25-30 cm, pH should be around 6 to 7.

Selection of Grass Species for Lawn Development

Some important grass species for lawn are

a) Bermuda grass (*Cynodon dactylon*) This grass is commonly referred as *doob* grass and also known as *hariali, arukampillu* and *durba* in vernacular. It is very commonly used for planting lawn due to its fast growth, hardiness, less water requirement and reponse to frequent mowing. This grass makes excellent turf and is tolerant to frost.

b) Korean grass (*Zoysia japonica*)This grass is suitable for smaller areas and home lawns. Korean grass grows slowly and there is no requirement of frequent mowing.

c) Kentucky blue grass (*Poa pratensis*) This grass is of very fine texture and soft and is like carpet. It grows best in full summer.

d) Rye grass (*Lolium perenne*) and crested dog's tail grass (*Cynosurus cristatus*)These are the cheaper species and not very fine.

e) Wood meadow grass (*Poa nemoralis*) This species is suitable for growing under shade.

f) Grass suitable for temperate region- *Agrostis tenuis* and *Festuca rubra* are the grass species suitable for temperate regions.

Methods of Planting: Different methods for establishing lawn are

a) *Seed sowing:* Prior to sowing, relatively dried surface area is scratched to a depth of 2.5 cm. with the help of a garden rake. Total area is divided into 200 to 300 m^2. Since the seeds of lawn grass are very light and fine so it is necessary to mix with double quantity of fine soil and seed is broadcasted @ 500g/200m^2 .Then ground is rolled lightly, watering is done, seed will germinate within 3 to 5 weeks period. When grass attains a height of 5 cm, it is clipped off.

b) *Dibbling:* After land is ready, well matured rooted or unrooted doob grass cutting is obtained from close cut lawn or nursery. Then grass is dibbled at 7-10 cm apart. After 5-7 weeks, cutting is done. Lawn is ready after 4 months.

c) *Turfing*: It is the quickest method, but its cost is prohibitive. Turf is a piece of earth of about 5 cm thickness with grass thickly grown on it. The piece may be square, round and free from weeds. It is laid on ground closely to each other in a bounded alternate pattern, like bricks in a wall.

d) *Turf plastering:* A paste is prepared by mixing garden soil, fresh cow dung, and water. Bits of chopped fresh grass roots and stem are mixed with this. The paste is spread uniformly on ground and covered by soil having 2 cm thickness. It is not suitable in dry and hot climates.

Maintenance of lawn: Lawn needs regular maintenance. Some important practices to be followed are

i) *Weeding*: All weeds should be removed with roots before flowering. Most serious weeds are Motha (*Cyperus rotundus*) and dudi. Frequency of weeding is more in rainy season compared to winter.

ii) *Rolling, mowing and sweeping*: Object of the rolling is to help grass to anchor itself securely and keep surface leveled. Mowing consists of cutting grasses at correct height with machine. Grasses should not be allowed to grow more than 5-6 cm in height. Sweeping of lawn is done thoroughly after each mowing to clean out grasses which might have fallen from mower box.

iii) *Irrigation*: Doob grass is shallow rooted and frequent light irrigations are necessary. The winter dew is important and each morning the dew should be brushed into the grass by drawing a hosepipe over the grass, before the dew evaporates.

iv) *Scrapping and raking*: Continuous rolling and mowing leads to formation of hardy crust. Therefore grass is scrapped at ground level with the help of *Khurpi* in month of April or May. Scrapping is followed by raking to break crust.

v) *Top dressing with compost and fertilizer*: After scrapping or raking a compost consisting of good garden soil, coarse sand and leaf mould in proportion of 1 : 2 : 1 is spread over lawn to a depth of 3-5 cm. For this, 100 kg compost is required for 100 m^2 area. Single Super Phosphate is applied @ 1kg/50 m^2. Compost is also used as top dressing in month of September – October and from October–April, Urea is applied once in every month @ 500gm/50m^2.

vi) *Frost injury*: This problems occurs in the northern plains that frost prone. This can be avoided to a great extent if grass is sprayed with water every evening and in early morning after frost.

Terrace

Terrace refers to a raised landform in a garden that emphaise its feel and visual interest. In planes, terrace effect is artificially created through soil filling and excavation while in hilly tracts it is not possible to have a large piece of land in one plane for laying a garden and hence gardens are laid in terraces, where it is a natural phenomenon. The land for gardening may not have any natural undulation for terracing in case of plain landform. It will be worth while to have a terracing to break the monotony and bring novelty into the garden. For a household garden one may remain content with making one terrace only.

In the plains preferably, the terrace should not be situated on the western side, as it will prevent the inhabitants from relaxing in the afternoon because of the hot sun. A terrace may also be constructed in front of a boundary wall or a dry wall. A terrace should form a natural link between the two and should give a full view of the garden. It is advisable to make the terrace about 45 cm above the general level of the garden supported by a retaining wall or bank. Ample space should be provided to utilize it as a sitting place. The major portion of the terrace may be paved with stone or brick, but before doing this, space should be left for small flower beds or for creepers to climb on the house walls. If the terrace is sufficiently wide, patches of lawn in between paving will add the beauty to a garden. In places having a hot summer, it is advisable to keep the paving to a minimum and leave more space for beds and grass to keep the place cool.

Children's Play Area

Play area for children may be planned in a spacious home garden or in public parks. Playing area in the midst of trees and garden is wonder magic for the over all development of child growth as reported in various research as well as by the Richard, the author of 'Lost child in woods'. Its important to have open centre and neatness in the children's area. Besides, some playing features like swings, slides or see-saw may also be installed in the play area. A raised plat form or terrace area is appropriate for kids. Its important to either have good paving where the kids can sit together, sing or play and eat and sand or loose gravel where swings and slides are installed. For eating purpose , wooden or stone seating with desk is appropriate when installed on raised platform or deck area. Besides, such areas will enclourage kids for creative ideas like drawing, painitng or playing music. Lawn that is highly hardy and tolerant to walking and playing may also be laid down provided it si timely trimmed and kept short. Children love to escape from adults but for safety point of view, it is necessary to have a eye on them from window or patio. Also if the space permits a secret tent or a tree house may be well installed. Bamboo canes and wood are appropriate for this purpose. The best swings are hung from stout tree branch. Trees with low and strong branches are defiantly a good climbing and playing site for children. Trees like *Ficus benghalensis*, *Saraca ashoka*, *Ginco biloba, mango, Plumeria etc* are perfect for this purpose. Such areas are magical corner for kids. Near annual beds and delicate plants should be avoided near the playing area. Also plants which are little acidic or even poisonous need to excluded from the garden. Hedges of hardy plants like *Clerodendron inermi*, golden duranta may be planted. Even small and shallow water body having small fishes may also be developed where children will learn and love to feed fishes.

Steps and Ramps

Steps and ramps in a garden help to break monotony of plain level and to climb over terraces or different plains. Steps can be gracious garden accent that set the mood for the entire garden landscaping scheme. Steps are comparatively efficient in space utilization while ramps score over when it comes to accessiblity for bicycling, baby strollers and wheel chairs. Steps provides sure footing equilibrium and sense of balance while passing through a level change in a better way then ramp. Also more vertical space can be covered through a smaller horizontal space as compared to ramp. Further, steps serve other functions like defining the limits of outdoor space, act as gateways or doors between adjoin g outdoor rooms and even provide casual sittin g surface. Aesthetically, steps also act as focal point or accent at the end of a walk, also for display of flowering pot plants, they create strong horizontal lines in outdoor spaces, provide visual fascination through abstract pattern of repeating lines.

The materials used for step are bricks, concrete, stone, wood or gravel. The gravel can be held in position with a stone-retaining wall. Grass should not be used in the steps as it becomes slippery when watered.

A flight of steps supplemented by plants, troughs of flowers with some special lighting appears more decorative. While deciding steps, tread (step), riser (vertical height), landing or platform (large level area between series of steps) should be considered. Steps should be wide with low riser. Steep and narrow steps are unsuitable. General rule is: twice the height of steps (riser) + the width of steps (tread) = the length of the human step i.e. 66 cm. For eg. Step height of 5.5" with 15" tread is suitable for gardens. The depth of the step should be minimum 11". Shadow line below step may be created by extending step or through an indentation, this makes the step more obvious and visible from distance. The maximum number of risers in a set of steps should not be more then 4′ height for unprotected steps, followed by landing. Platforms or landings break the visual and psychological monotony of a group of steps and makes the elevation change easier to traverse. It also acts as pause or resting plain in between steps. A variety of material can be used for making steps in garden like concrete, exposed aggregate concrete, mortared bricks, tiles, railroad tie risers with concrete treads, wooden boards, wooden boards with pipe supports etc.

For ramp, the degree of floor slope should not exceed 8.33%. Thus, ramp would need 36′ horizontal distance to cover 3' height while similar distance can be covered with 6′ distance through steps. The sides of ramps should have at least 6" curb. The maximum horizontal distance covered should be 30′ for ramp followed by landining. The ramp should appear as a well coordinated element within the total design scheme. The location of ramp should be on primary, direct lines of movement. Steps and ramp can also be coordinated to g ether tactfully.

Cheeekwall and handrails are also important design and safety aspects in steps and ramps. Cheekwalls visually end the steps, act as retaining walls to hold back the slope from the steps and help sure footing of pedestrians. Cheekwall may be gradient on top like steps or straight. Handrail provides support to hold on for the climbers on the steps. Handrails may be located on top of or along the inside of cheekwalls, depending upon the height of the cheekwall and the appearance desired. Handrails located at 32-36′ height from the riser and should extent about 18" horizontally beyond the nose of top and bottom treads for holding from distance as comport and support. For wider steps, handrails should be located at 20-30′ intervals across the width of the steps.

Pavements

Pavement is one of the strikingly hard structural feature of a landscape design that is relatively fixed and non-changing and multifunctional. It is a hard natural or artificial material placed or laid on a ground plane of outdoor space to establish a durable surface whole satisfying design objectives. Pavement is a good means of emphasizing unity within all the elements of landscape. Pavement function as neutral setting to buildings, sculpture, potted plants, displays, benches, swings etc

It is relatively expensive compared to lawn or other vegetative ground covers but is less maintenance requiring with little wear and tear and more functional uses. The functional use of pavement include its intense use for movement either by walking, cycling, baby tramps and wheelchairs as well for taking a break and having rest. It provides clean environment by avoiding wind erosion and dust. Careful placement of paving provide direction and indicate movement when used as a thin strip or in a linear manner on ground like a walk leading to a building, paths in park and across college campuses. The width and shape of pavment also influence the rate and rhythm of movement, where wider paving suggests casual movement while thinner paving stimulates careful and rapid movement. One important function of paving is creation of sense of repose and rest, specially when paving is done in a large area with non directional forms and patterns. Further, changes in pavement colour, texture, pattern and material cn be used for identifying different purposes like for path, arrival area, resting or sitting area, dinning place, gathering, focal area or playing area. Pavement also influence scale through the block size, orientation and spacing. Different pavement material and patters create and reinforce diverse spatial feelings as refined, rugged, quiet, aggressive, rural or urban. Besides, pavement can serve as strong feature to provide visual interest. A distinct paving pattern creates special point of appeal and a strong sense of place. Considerations while designing pavement are a) avoid too much variation in pavement patterns within a design and one pavment should dominate the whole design, b) the material used should be visually and functionally integrated into the entire design, c) should be level change and to demark to pavement types, d)smooth textured material for pavement should be selected for space emphasis to imply quiet and peaceful effect. Numerous pavement material is in three groups 1) Loose pavement such as gravel 2) Unit pavers like brick, tile, concrete pavers, stone *etc.* 3) Adhesive pavement with a binding agent in cement concrete or asphalt. Gravel's loose textures quality allows it to adopt any shape or form to which it is applied and to be appropriately used in informal and rural settings to portray naturalistic effect and to provide textural interest and contrast on the ground plane. Stone may be sedimentary like sandstone and limestone, metamorphic like marble and igneous like granite. Further, based on geographical location, stones are classified as fieldstone (irregular shape), riverstone (rounded), cobblestone (rounded and flat),flagstone (layered and thin slabs) and cut stone(that can be chiseled easily). Stones look highly elegant naturalistic as well as modern depending upon the selection. Concrete pavers or interlocking concrete blocks or pavers, as well as stamped and grouted concrete also provide a nice clean look. Bricks can be arranged in different patters like stack bond, running bond, basket weave, herringbone, basket weave and stack bond and basket weave and variation. Tiles are another smoother option for paving. Concrete along with grass paving provides a softer appeal to the hard paving structure.

Seating

Seating in the form of benches, walls, swings, planters, chairs is an important feature in the landscape that directly influence comfort and enjoyment of an exterior space. Seating arrangements in the landscape serves to provide feeling of rest and enjoy fresh breathe among beautiful flowers and plants and also function as waiting convenience. Garden seats are a necessity in any garden for resting and peacefully enjoying the beauty of a garden in the soothing environment. Particularly home gardens require seats where family members visit frequently. For this reason, it should be comfortable and durable along with artistic look and should merge with the surroundings. Relaxed chair cum sleeping chair is perfect resting, enjoying sun simply lying down. Seating also serve as a location for conversation among a group of people depending upon its arrangement. It also serves as a point to watch the rest of world i.e children playing by, people walking and butterflies flying by specially when located near walkways, busy corner or overlooking a plaza. Garden is the best place for unlocking creative abilities, where seating arrangement in the garden encourages book reading, poetry writing as well casually eating or enjoying winter sun. Important consideration for locating and designing seats in a garden include a) location should be a protected corner or beneath shady trees or shrub or near pond or pool side, c) may have overhead screen or shelter to enjoy rainy weather, d) should be integral part in the overall design, e)should provide a sense of privacy and security f) layout should be as per the functional use like opposite or cornered for conversation purpose or meeting purpose. Regarding the dimensions, the seating height should around 18" with 12-18" width and 15"back if provided. Sitting arrangement on a raised platform and hard surface or pavement is considered better. Sitting arrangement could be linear seat or cornered seat or modular seating. Modular seating layout permits individual or group seating as well as facilitates conversation and orientation selected by the users. Garden seating could be fixed or moveable. Simplicity and comfort is invariably the key to good design of seating. Fixed seating is generally for patios and raised decks. A tree seat also looks appealing. Various material are being used for making seating like wood, stone, iron, bamboo cane, wooden logs, bricks or concrete or even plastic. Wrought iron and genuine cast although look elegant are expensive, where less expensive alloy imitations are available which are light whitened also. White paints look although graceful for these seating but may catch more dirt, where green is also smart version. Wooden benches should be either varnished or kept neutral and natural with wood colour paints. Plastic seating is advisable for patio for viewing garden.

Wooden and fabric seats are comfortable and compliment the garden. Wooden seats should be made of good timber to withstand the changing weather. The chair or bench should have an appropriate design. Iron or stone or concrete seats are more durable but get easily heated in summer and become cool in winter, which makes them less comfortable.

Carpet Bedding

Carpet bedding refers to covering an area, preferably a bed or a series of beds, with dense, low-growing herbaceous plants, either flowering or foliage plants, according to a set design. It is a common feature of formal garden. Lawn is the best example of carpet bedding. In such designs, often a figure or some letters are cut out with the help of plants having different growth habits or having different coloured leaves.

In a carpet bed, the plants are planted closely according to a set design. The plants are trimmed regularly to keep them within the limits of the design, and gap filling is done as and when required. Once in every 3-4 months the beds should be top dressed with well-decomposed manure. *Alternanthera, Aerva, Eupatorium, Golden duranta Cineraria maritime, Coleus, Echeveria, Pilea muscoca, Portulaca, Sedum rupestre, Torenia asiatica,* etc. are recommended for carpet bedding.

Flower Beds

The display of flowers in the best possible way is the most essential criterion of a flower bed. Flowers look best when massed in a bed. The flower beds can be planted with winter flowering annuals sown during September/October or summer flowering bed can be planted during February/March and for rainy season flowers, seeds are sown during April-June. Flower beds should be simple in design such as circles, rectangles, squares and ellipses as it is easier to maintain them. Kidney or heart shaped beds in the midst of a lawn also look very attractive.

Flower beds are most important part of a formal garden. In an informal landscape garden, the number of flower beds is less compared to a formal garden, as the flowers are mostly grown in the borders or in front of the shrubbery, for giving live ness to the landscape design. The beds should always be placed in full sun except when these are meant for growing shade loving plants. Rectangular beds are best placed near the building to harmonize with the straight lines of the house. While arranging the beds in a lawn, the grass verges or paths in between beds should not be less than 45-60 cm, otherwise it will be difficult to run a lawn mower.

Annual flowering plants can be used with single or more colours together with proper designing. Time and season of flowering is also an important criterion. Plants for flower beds are *Ageratum, Alyssum,* Antirrhinum, *Calendula officinalis, Celocia, Chrysanthemum carinatum, Coreopsis, Cosmos, Dianthus* sp. (Carnation, Pink, Sweet William), *Dimorphotheca aurantiaca, Gomphrena, Iberis umbellata, Impetiens balsamina, Phlox, Salvia, Tegetes, Verbena, Zinnia, etc.*

Shrubbery

This is also an essential feature of garden. Shrubs are of permanent nature and once planted will become a permanent feature unlike the seasonal flowers which are to be replaced every season. From planting shrubs in shrubbery point of view, shrubs are classified as

i) *Shrubs having ornamental foliage and attractive foliage, eg:* Acalypha, Duranta, Variegated Hibiscus, Pedilanthus, *etc.*

ii) *Flowering shrubs, eg:* Pentas, Ixora, Tabernaemontana, Hibiscus, Mussaenda, *Cassia glauca,* Caesalpinia, Callindra, *Bauhinia purpurea, Hemelia patens, Jatropha rosea,* Plumbago, *Meyenia erecta, etc.*

iii) *Shrubs which bear berries, eg: Braya ebens, Duranta, Rauwolfia serpentine, Carissa caronda, Poincirus trifoliate, Ochna kirkii, O. squarrosa, etc.* They are further grouped according to their requirement of sunlight.The shrubs in a shrubbery will provide flowers throughout the year. The beds meant for shrubberies need not be very formal but should appear more natural. Shrubbery can very effectively be used, for hiding one portion of a garden from the other, besides adding beauty.

Borders

Beds which are more in length than breadth and contain of a heterogeneous character are known as borders. There are three different types of borders, namely, herbaceous, border, annual mixed border and mixed border.

Herbaceous border: It contains hardy perennial herbaceous plants which die down to ground level after flowering, but put up new growth from the roots in the next season. Herbaceous border needs a good depth of soil and a sunny situation. The border may be placed against a wall, a fence, shrubs, a hedge or form a double border divided by a grass path. In an informal design the border should have curved margins. Some characteristics of herbaceous boarders are:

a) Frontage:-Best frontage is a stretch of neatly kept lawn because this provides a good foreground colour. A paved brick or stone walk is considered the best because this provides a firm and clean walk throughout the year.

b) Height and arrangement: In a wide border, plants of all heights are accommodated. The grower has to give enough attention on the staking of tallest plants. The maximum height should be limited to 3 m.

c) Grouping and colour: In an average sized border each group will consist of 4-5 plants, while in large borders it may vary from 5 to 7 plants. Each group should be planted in drifts of irregular shapes than being put in box or round shaped drift as this may look patchy. Though generally the gradation in a herbaceous border is from dwarf to taller placed towards the back, it can be broken here and there to avoid monotony or too formal look.

Plant selection: The main aim of herbaceous border is to have colour in the garden throughout the year and to exhibit the proper plant material in a most artistic way. For year round flowering, plants are compiled in such way that in every season some of these are in flowering. Some examples are listed here which are commonly used for herbaceous borders are *Amaryllis* (different cultivars), *Asystasia* species, *Beloperone guttata, Canna* different cultivars, *Clerodendron balfourii, Chrysanthemum, Coreopsis grandiflora, Crossandra* different species, *Eranthemum, Gaillardia perennis, Lantana sellowiana, Michaelmas Daisies, Plumbago copensis , Portulaca*, Perennial *Rudbeckia, Ruellia* different speciea, *Russelia juncea* and *R. floribunda, Salvia, Solidago canadensis*, Perennial *Verbena, Vinca* different species and *Zephyranthes* in different colours, *Dianthus chinensis, Sedum spectabile, Statice latifolia, Viola cornuta, Delphinium belladonna* (different cultivars), *Gypsophila elegans, Lupins, Phlox decussata, Tradescantia Helianthus annuus, Hollyhocks* and *Peony.*

Plants for mixed border: A mixed border generally includes the following type of plants:

i) Herbaceous shrubs (i.e. not woody) with less spreading root system,

ii) Sub-shrubs which are of dwarf growth and have soft stems,

iii) Herbaceous perennials,

iv) Bulbous plants (dahlia, tuberose, zephyranthes, amaryllis, etc.) and

v) Annual flowers. A list of shrubs, sub-shrubs, herbaceous plants, bulbous plants and annuals classified on the basis of height, for mixed border, is given as below:

a) Tall (height 1-2 m): *Acalypha* different species with ornamental foliage, *Caesalpinia pulcherrima, Cestrum aurantiacum, Dahlia* tall cultivars, *Gardenia florida, Hibiscus mutabilis* and other tall varieties, Hollyhock, Sunflower, Peony, Dahlia, heliconia rostrata, H. psittacorum, *Ixora singaporensis, Largestroemia indica, Nerium oleander, Poinsettia pulcherrima, Tecoma capensis, T.stans, T. gaudichauri* and *Thunbergia erecta.*

b) Medium (0.5 to 1 m): African marigold (tall cultivars), *Antirrhinum majus, Beloperone amherstiae, B. guttata*, Canna medium cultivars, *Chrysanthemum* tall cultivars, *Coreopsis grandiflora, C. tinctoria, Gaillardia, Galphimia gracilis, Gladiolus, Phlox decussata, Plumbago capensis* and different species of *Salvia, Zinnia elegans.*

c) Dwarf (10 to 50 cm): *Ageratum* different species, *Alternanthera,* China aster different cultivars, Candytuft, *Coreopsis drummondii, Dianthus chinensis*, Gerbera, *Lantana sellowiana, Michaelmas daisy, Phlox drummondii, Portulaca, Rudbeckia* (dwarf), *Salvia coccinea, Verbena hybrida, Zephyranthes* and *Zinnia linearis.*

Topiary

Topiary is the art of training and pruning plants to provide a particular shape. This when carefully done adds a feature of fun and attraction in garden. It has special importance in formal and cottage gardens. Topiary could simply be a battlemented hedge or geometrical shapes (sphere, cones, pyramids, cooumnar, etc) or imaginative or wild shape or in the shape of birds or animals. Most hedging plants can be used for topiary, although yew, *Lonicera nitida, golden duranta, Ficus benjamina, Clerodendron inermi, Sophora tomentosa, Putranjiva roxburghi, casuarinas, bougainvillea, Hibiscus rosasinenesis, Taxus baccata, Vernonia elaegnifolia*, etc are widely used for making topiary. Simple forms like cones or a waved hedges can be developed through careful pruning. However, for more precision requiring shapes like that of some abstract, sphere, birds or animals, a metallic structure or a wired structure of that particular shape is created and plant is trained with its by trimming any branches exposing outside it. Initially, the first leader of the shrubs are tied and trained round the wired structure then followed by trimming of laterals going out of the way or shape. Once the shape is established, regular pruning is must to avoid any clutter or uneven shape. Hand shears work well for topiary work, however, mechanical hedge trimmers my be used but with careful handling as little mishandling may damage the whole shape. Care should be given for providing fertilizers, water and even mulch to the topiaries.

Garden Drives and Paths

A garden drive or path provide direction and feasibility to garden, adjoining buildings and to various elements and components within the garden. A good garden drive or path should harmonize with the garden and provide a flat, dry and aesthetic passage for the people and persons or vehicles. A garden path is generally made up of gravel, paving-stones, crazy paving bricks, or grass while.

i) *Drive:* Garden drives should be a little higher (2.5 cm) than the surrounding ground and its width should be about 3-5 m. Garden drive is made up of gravel, stone or concrete as it is meant for light weighted vehicles also. In most of the localities a gravel drive is quite useful for day to day work. A gravel drive should be given adequate foundation to prevent it from sinking. The soil is excavated to a depth of about 30 cm but in heavy clayey soil this should be a little more to ensure proper drainage. The surface should be a little higher at the center and sloping to the sides. The edge of the lawn, if it encroaches onto the road, is cut straight with an edging iron. The drive should be kept free of weeds with the help of some weedicide. Asphalt or concrete drive provide formal look with more stable perspective. A layer of 5 cm of concrete is sufficient for the garden drives

ii) *Paths:* Garden path is for walking within the garden and enjoying its different components and features. The width may vary as per the choice, however, it should atleast 60 cm wide and preferably be between 90 and 120 cm. A wider path allows easy passage for wheel barrows and other garden implements and will not be obstructed by plant growth. Plants should not be generally too high or too low from the adjoining ground except in a marsh or a rock garden. Stones, bricks or tiles or cement blocks are generally used for path. Gravel path is made in the same way as gravel drives except that the excavation and foundation material will be less. The strong red bricks may also be chosen for paths that may either be laid flat or on their edge. Many patterns can be woven such as herring-bone or basket patterns. The bricks in the paths are arranged in such a way that the cross-joints are not continuous. Paths of stone paving are more popular, especially where stone slabs are easily available, as these are permanent and impart an air of quality. Crazy Paving comprising of stones of different sizes and irregular shapes look the most attractive in informal setup. Stones for crazy paving are generally thinner (2-3.5 cm) than those used for stone paths. To make it permanent, the paving should be bedded over 2.5 cm cement mortar. Even grass paths look most desirable and attractive specially for walking bare foot. Paving stones may be placed at 60 cm intervals in single row just below the ground level on grass path. This will give a stepping-stone effect. Alternatively, groups of two or three bricks or wood logs may also be used. The best way to make a grass path is in the form of curves. Otherwise, the grass path is to be started from seed of a hardy grass.

Garden Lights

Garden lights are essential in garden to serve important purposes like safety, security and decoration. It is essential in outdoor space like slope areas, steps as well as sitting area, entries, curbside areas, etc to get rid off any danger and to get a feeling of safety. Lighting in a garden can give a completely different appearance to plants and features such as water or statues. Lighting at night emphasize the usable living area stimulate the mood and emotion of the visitors, or showing up. It allows the best features to be lighted while leaving others hidden and brings the garden to life in a totally different way. Lightening along with water use and music effect captivates everyone's soul in a garden. Artistic electric candles on tables, lamps on lamps posts or dancing light arranged on plants are various ways of lighting in a garden. Modern lights designed in varied artistic mode are available for garden purpose. Electric wires should always be underground and safe in a garden. Further, garden lightening allow to create many decorative effects after dark, by virtue of patterns of light and shadow produced by night lighting. Varied lightning effect can be produced by different source of lighting like down lightning i.e the source of light is above the subject like overhead lamps or light mounted on trees or uplighting i.e projecting lights upwards making the foliage glow. Uplighting produced intricate light and shadow effects when combined with downlighting. Water illlumination also provides dramatic effect by a simple white spot light playing on moving water. Even underwater lighting is becoming popular by using special sealed lamps designed to be submerged or to float. Before, fixing the light in the garden, one should move the beam of light around to see which direction or angles looks more stunning. Besides, all important areas like path, water feature, sitting area or dinning area and the open space should be well illuminated for practical purpose. Light modification is also important consideration for garden like accent lighting to focus on particular focal point or area or diffused lighting i.e light filtered through translucent materials such as plastic and canvas to obtain glare free illumination. Halogen light although bright for out doors, do not spread and diffuse and may cause glare problem in garden. The incandescent (240 v spot lights) are highly brilliant but need to be installed at proper angle only otherwise is too high bright and may be influence the quiet and silent garden effect. Light intensity can also selected from 120 volts to 12 volts as per the need. 120 volts light system is the best choice for situation requiring brilliant illumination but need proper installation system and proper location to be installed. Passive infra red (PIR) lights are also available which glow automatically when anyone approaches the house, which are good from safety point of view specially when you are not in the garden. 12 volt low voltage lighting is easy, available in may type of fixtures and can be installed are various location in the garden and may be used in producing dramatic and stunning effects. For softer illumination and for smaller areas as well as well as for back lighting, low voltage system can work well.

Garden Decoratives or Adornments

Garden decorative or adornments add the charm of extra care and purpose with comfort. Garden ornaments are an integral part of a garden, and express the basic philosophy of the garden, i.e. to serve as for aesthetic appeal, contemplation and pleasure. Some of the garden ornaments are described below:

Sculptures and Statues

Sculptures objects demands high aesthetic sense for choosing the right kind and right position and spaceous location. Statues of animals or inanimate objects are used in the garden to improve the over all look of the garden. Its important that the statues chosen for the garden should blend well with garden style and feel. Too many statues should be avoided as they provide a gallery look instead of garden. Sculpture often gives rise to passionate responses when viewed, hence, the size, form and material should be chosen artistically. Contemporary art provokes a subjective response which can be successfully exploited in the garden in the form of scultures and statues. Classic figures can augment the garden's focal point when placed at the right position. Human figures look great in an alcove or on a plinth in a dull corner but sometimes may raise controversy. These figures can be accentuated with surrounding plants. Animal figures may be set among plant or in lawn. Abstract ornaments make considerable impact so need to placed with restraint. The statues can be carved out of stone or made of bronze. The statues can be placed in the midst of running water like in stream or in a pool or at the intersection of two roads or at the end of the road or near the doorway of the house.

Fountains

Fountains are one of the main attractions in the garden which attract people of all ages. Various designs in fountains are available. There may be a straight water jet or a number of finer water jets converging in the form of an umbrella or rainbow. To make the fountains more colourful, colourful lights are provided under water with automatic adjustment of colour changes at regular intervals.

It can also be linked with the rhythm of music to make it more appealing and soothing. Different type of fountain creates various types of sound.

Floral Clocks

Floral clocks are wonderful and highly attractive feature specially in public parks and famous formal gardens. Huge clocks are placed on gently sloped ground which is electrically operated. The machinery of the clock is under the ground in a chamber with only the hands showing above the ground against carpet bedding plants or flower beds. For eg, a carpet bedding of green plants like *Pilea muscosa, Sedum*, lawn, etc. with the figures made of red plants like *Alternanthera, Areva*, etc. or *vice-versa* which looks quite ornamental. Instead of live plants the dial can be decorated with various coloured pebbles or pavings also. Some famous floral clocks include that of Niagra fall park, Canada, etc.

Bird Baths

Bird baths are highly appealing garden adornment as they attact birds, add the bio-aesthetic look as well as provide interesting mobility in the garden. This feature is not only appropriate for public gardens but even for private and home gardens. Further, it is better choice than sundial specially when placed in shady location. Bird bath consists of a large shaped bowl like containers, made up of concrete or wood or plastic and fixed over a pillar or column. It is placed at one meter height. Water is stored in this bowl for birds to drink and bath. It is generally placed in the corner of the garden and near the trees. Berry bearing plants if planted nearby attract more birds. Besides, eatable for birds that is multi-grains may also be provided for the birds to feed along with water. A beautiful wooden bird house like structure may also be attached on nearby attached column, where the bird would sit, rest, play. This will encourage more birds visiting the garden and even building their nests on the nearby trees. A birdbath also looks elegant when placed near flower bed or even on deck or terrace or patio area that facilitates watching birds bathing with garden viewing.

Ornamental Tubs, Urns, Vases and Plant Stands

Plants displayed in ornamental tubs or urns at suitable places look beautiful by adding beauty to garden. The urns or tubs can be made of timber, concrete, ceramic, fibre, metal but preferably of bricks or carved out of stone. These are generally fixed over ornamental pillars for permanent setting. If there is paved path, an ornamental vase can be placed at the end of it or one each on both the sides. These can also be placed near the gate or staircase of the main entrance. Ornamental urns preferably of metal with outside carving look beautiful in the terrace, near the steps or even inside the house. Suitable ornamental plants should be grown in such containers.

Plant stands are very useful structure for emphasizing the beautiful pot plants. These are generally made up of mild steel rods molded in different fashions with various sized rings attached to hold the pots. These are very useful when placed in roof or balcony gardens. They form a good display in terrace garden, at the entrance of the house or any other place in the garden. Plant stands, specially designed for fixing in the walls of the house are also available where indoor plants can be displayed but these should be painted with complementary colours to harmonize with the background. Specially designed plant stands are also available for creating of vertical garden concept.

Sundials

A sundial in garden may be used as a focal point, a centre-piece of a flower bed, can be placed in the centre or at the end of the lawn, it indicates the end or terminal point of path when placed at the end of the path as well as a good feature in a sunken garden. The sundial should be positioned in a place where the shadow from a tree or building does not fall for a long duration. The column of the sundial should be made of bricks, tiles, stones or combination of all. The top is generally square or circle where the sundial with the compass is fixed. This is made of stone or concrete or metal. The height of the sundial should not exceed 60-90 cm to look easily. Live sundials are also made up of live plants complete with figures, compass, etc. in the form of topiary. The position of sun dial needs to be a well sunny location. The best place for sundial is a centerpiece for a formal garden. Centre of the Lawn area is also ideal provided if the area is large enough, however, for small areas, a sunny corner is also appropriate for locating the sundial.

Japanese Lanterns and Stones

Japanese lanterns are low and decorative and preferably carved form in stone. The column may be cylindrical or square faced. The fire box should also be carved. The roof may be broad-roofed, commonly called “Snow View”. Generally these type of lanterns are made of white stones or marble or decorative metals. Lanterns are placed near the pool or are fixed over ornamental pillars near the house at entrance.

Various size and shape of stones can be used alone or in a group as viewing point of interest. Some of the mall trees or shrubs can be displayed in the background. Stones can be displayed in lawn or corner of the garden on the small hillocks. Some time it can also serve as a natural seating place. These are sometimes grouped for edging purpose or there is important placement of stones in rock garden and also for pond edging.

Pillars

Pillars in gardens are an effective way to accentuate vertical feel in a garden. When trained with beautiful flowering climbers that add visual point of interest. The pillar could be temporary or permanent feature. If used for annual climbers like nastrutiums, star ipomea, these can be made up of wood or bamboo canes and are temporary feature. Iron or galvanized iron poles are long lasting and can be efficiently used for more perennial climbers like rose, jasmine, clerodendron, Adenocalyma, Petrea, Echites, etc. Along with the pillars, it is necessary to wind wire netting to support the creepers. There is need that the base of pillar to be well erected in 1-2' deep hole, with well and firm filled soil. Further, for a more permanent structure, concrete or gravel may be filled. Location of pillars should be near the entrance or to accentuate a focal point in the garden or even at the sideways of the path to accentuate it. A fragrant flowering pillar path adds romantic charm and secluded look to the garden.

Fence

Wall

Screen

Gates

Plate 39

Raised Bed

Hedge

Edge

Plate 40

Pergola

Overhead

Lawn

Plate 41

Steps and Ramps

Children's Play Area

Pavement

Plate 42

Seating

Topiary

Plate 43

Paths and Drive

Garden Light

Statues

Plate 44

Fountain

Urns and vases

Bridge

Plate 45

Colour Plates

Plate 1: Landscape Gardening: Importance and Benefits

Plate 2: Landscape Gardening: Importance and Benefits

Formal Garden

Informal Garden

Plate 3

Informal Garden

Wild Garden

Plate 4

Family Garden

Secluded Garden

Terrace Garden

Plate 5

Paved Garden

Rock Garden

Plate 6

Moon Garden

Bog Garden

Sacred Garden

Plate 7

Home Garden

Kitchen Garden

Plate 8

Balance

Focal Point

Space

Plate 9

Axis

Unity

Plate 10

Repetition

Rhythm

Formal Landscaping

Informal Landscaping Style

Plate 11

Variety

Plate 12

Characteristics of Plant Material

Plate 13

Divisional Lines

Plate 14

Mobility

Plate 15

Time and Light

Plate 16

I Landform

Level landform

Convex Landform

Ridges

Valley

Valley

Plate 17

II Plant Material

Acroclinum

Altea rosea

Antirrhinum majus

Annual sunflower

Arctotis

Californian Poppy

Briza

Plate 18

Bromus

Calendula

Candytuft

Chrysanthemum coranarium

Coreopsis

Dahlia

Dianthus barbatus

Dimorphotheca

Euonymus fortuneii

Plate 19

Gamelopis

Gazenia splendens

Geranium

Helichrysum

Hyacinth

Lagurus

Limonium

Melucella Leavis

Marigold

Plate 20

Mesembranthemum

Nasturtium

Nigella

Ornamental cabbage

Pansy

Pimpinellifolium

Poinsettia

Shirley Poppy

Plate 21

Primula malacoides

Ranunculus

Salvia

Sweet Allysum

Sweet Willium

Plate 22

Bulbous plants

Agapanthus

Amaryllis

Clivia miniata

Day lily

Deffodil

Spider lily

Lilium

Plate 23

Cactus and Succulents

Aeonium arboreum

Agave schibigera

Begonia rex

Dizygotheca elegagantissima

Echeveria peacockii

Euphorbia Thai hybrid

Fuchsia sp

Furcraea

Plate 24

Gynura aurantiaca

Hawothia fasciata

Hedera helix

Hyacinthus

Kalanchoe tomentos

Lampranthus roseus

Neoregelie

Opuntia microdasys

Pereskia acculata

Portulacaria

Plate 25

Cycus

Cycas

Zamia furfuracea

Ferns

Nephralepis exelata

Nephrolepsis biserrata

Sedum morganianum

Sedum rubrotinctum

Senecio rowleanus

Senecio serpens

Plate 26

Tradescantia sillamontana

Trees

Adansonia digitata

Alstonia scholaris

Amhrestia nobilis

Araucaria

Butea monosperma

Bauhinia pupurea

Beech tree

Plate 27

Callistphus chinensis

Calophyllum inophyllum

Cassia fistula

Cassia nodosa

Cassia renigera

Couropita guinensis

Cochlospermum gossypium

Delonix regia

Erythrina crista-galli

Plate 28

Jacaranda

Schleichera oleosa

Magnolia

Michekia chamaca

Mimusops elengi

Nerium oleander

Peltoforum

Pisonia alba

Plumaria

Plate 29

Plumera

Polyalthia longifolia

Prunus

Pterospermum acerifolium

Salix babylonica

Samanea saman

Sarrica indica

Spathodea campanulata

Plate 30

Tabebuia

Tabebuia heterophylla

Acalypha

Adenium

Aglonema

Allamanda

Alocacia

Aralia

Bougainvillea

Canna

Plate 31

Clerodendron thomsonae

Croton

Dombeya

Dracaena

Eranthemum

Euphorbia leucosephala

Excoecaria bicolour

Hamelia Patens

Plate 32

Hibiscus

Ixora

Lantana camera

Maranta

Mussaenda

Nerium

Pachystachys

Phyllanthus nivosus

Plate 33

Plumbago ovata

Poinsettia

Schefflera

Syngonium

Tabernaemontana

Tecoma stans

Plate 34

Climbers

Adenocallyma allyaceanum

Antigonon leptopus

Thunbergia grandiglora

Vernonia

Wisteria chinensis

Combratum comosum

Quisqualis indica

Jaquomontia violacea

Peparomia

Plate 35

Petrea volubilis

Stigmophyllon ciliatum

Hiptage benghalensis

Tecoma grandiflora

Thunbergia alata

Herbaceous Perennial

Begonia rex

Geranium

Plate 36

Periwinkle

Mirabilis jalapa

Coleus

Kalanchoe

Maichalman Daisy

Plate 37

III Water

Flowing

Stastic

Plate 38

Fence

Wall

Screen

Gates

Plate 39

Raised Bed

Hedge

Edge

Plate 40

Pergola

Overhead

Lawn

Plate 41

Steps and Ramps

Children's Play Area

Pavement

Plate 42

Seating

Topiary

Plate 43

Paths and Drive

Garden Light

Statues

Plate 44

Fountain

Urns and vases

Bridge

Plate 45